𝒥 dedicate this book to my sister, Crystal, my lifelong best friend, and, in my opinion, the *quintessential* example of beauty and brains. From reading *Gone with the Wind* at the age of eleven to scoring near perfect on her LSATs, she's always managed to add more than a touch of glamour to her scholarly activities. After all, she's the only woman I've ever known to regularly run through the snow in pearls and four-inch heels, on her way to class ... across the campus of Harvard Law School. I love you, Crystal!

Praise For Danica McKellar and *Math Doesn't Suck*

"This groundbreaking book is just what this country needs: a fun and accessible resource to help spark undiscovered math abilities in girls, and to inspire the next generation of female scientists, mathematicians, and astronauts."

—Dr. Sally Ride, first American woman in space

"McKellar is probably the only person on prime-time television who moonlights as a cyberspace math tutor." —The New York Times

"At last, a math book that every girl will want to have. Think *Clueless* (the movie) meets Euclid (the famous Greek math teacher) . . . If you are having any trouble with middle school math, this book is your life saver. Buy it, keep it with you, and show it off."

—Keith Devlin, PhD; NPR's "Math Guy"; coauthor of
The Numbers Behind NUMB3RS

"[McKellar is] a terrific role model."

—Francis Fennell, president, National Council
of Teachers of Mathematics (as quoted in *Newsweek*)

"*Math Doesn't Suck* is much more than a math tutorial—Danica teaches the value of confidence that comes from feeling smart. This wonderful book is a roadmap to true success for girls, no matter what career they choose."

—Leeza Gibbons, national radio and talk show host

"Danica is a perfect role model. . . . I hope all parents of middle school students will give a copy of this book to their kids—it'll go a long way towards improving the math education in our country."

—Tony F. Chan, PhD, assistant director for Math and
Physical Sciences, National Science Foundation

"A brilliant and successful effort to bring a little glamour to the teaching of mathematics."

—Veeravalli S. Varadarajan, PhD, professor of Mathematics, UCLA

"This book sparkles with proven techniques for succeeding in math as well as in life as a confident young woman. It should be required reading for every girl aged 9 to 16!"

—Bettie B. Youngs, PhD, coauthor of the Taste Berries for Teens series

"*Math Doesn't Suck* conveys math in an immediately accessible and delightful way to young girls. My three nieces won't even suspect that I'm trying to turn them into mathematicians."

—Jonathan D. Farley, D.Phil., professor of Mathematics, CalTech;
consultant for the hit TV shows *NUMB3RS* and *Medium*

Math Doesn't Suck in Action!
A Sampling of Danica's Emails

I absolutely love your book. Before I read it, I constantly got Cs and Bs in math, but after reading it, I'm getting As for the first time in my life! This is an incredible book, and I think every girl should read it.

—Thorie, age 11, California

I am going into 7th grade this upcoming school year. About a week after your book went on sale, my dad bought it for me. I read it, and I loved it. As I was reading, it was like a person my age (but a genius) was explaining so much to me! I used to absolutely *hate* math, but now I love it, and I'm doing *so* well! I would totally recommend this book to any girl my age.

—Caroline, age 13, Virginia

Last year, in the sixth grade, I got a B+ in math—I know it's not really that bad, but I was disappointed that I didn't do better. I just didn't get math. Then on the first day of 7th grade, my mom handed me your book. After reading some of it, we had our first math test . . . and I got 104%! I WAS SO EXCITED! The night before, I studied from my textbook, but then I spent an extra half hour looking over the explanations in your book. That was what really helped! I am proud to say that math is now my favorite subject.

—Ally, age 12, California

I am a 13 (going on 14) year old girl who has ALWAYS suffered terribly compared to my dyslexic peers. After reading your book, though, I don't just succeed in math—I LOVE it!

—Alex, age 13, Georgia

Omg. I haven't even gotten to the third chapter in your book, and it's helped me understand math so much better. Love your book and love your sense of style.

—Nadine, age 11, New York

Since I read your book, I've been thinking about someone who inspires me—and I discovered that this person is YOU! I love your book sooo much. My favorite part is the "Prime Numbers and Monkeys" section. When I read that chapter, I laughed sooo hard.

—Kaylynn, age 11, Oregon

I'm currently reading your book—it's so interesting and funny! I didn't do so well in math last semester because I didn't understand everything, but since reading your book, it all makes sense—and I just got 100% on my last math test! Thank you!   —Isabel, 11, Georgia

Your writing is perfectly geared toward middle school girls. My 12-year-old daughter started reading your book as soon as she finished her homework yesterday afternoon. This morning, she took the book to school to show her math teacher. Last night she kept saying "I never really understood some of these things before, and now I really do!" This book could not have come at a better time. You have made an amazingly positive impact on us!

—Laura Catherwood, North Carolina

I am a high school science and math teacher in California, and I have been so frustrated with the sad state of enthusiasm for math in some of my classes. Your book is truly a miracle for me. Not only are most of the girls in my class enjoying it, but several of the boys are able to connect to your method of exposition. I just wanted to let you know how many minds and hearts you have truly changed because of your wonderful book.

—Robert Vanderschraaf, Physics and Physical Science Instructor,
Buena Park High School, California

Thank you so much for writing this book. I bought two copies and already three of my stressed-out Honors girls have read it. They are now in Algebra (8th grade) and one said if she had read the book last year she wouldn't have cried so much. Again—thank you!

—Deborah Boutier, Emma Ousley Junior High School,
Arlington, Texas

I just purchased your book for my 11-year-old daughter. I think you've hit the nail on the head with the audience you're targeting; you've done a great job making exciting and interesting a topic that can at times be dry. My daughter and I spent a while reading your book last night, and she had a smile on her face the whole time.

—Ed Rosenthal, Simsbury, Connecticut

For the past decade, I have taught a course for prospective elementary and middle school math teachers. The next time I teach this course, I plan to require *Math Doesn't Suck* as supplementary reading. This lively book features many new teaching techniques and useful step-by-step procedures. Danica's explanations are clear, thought provoking, and convincing. I read the book word for word and enjoyed it to the end. I recommend it to all prospective and in-service mathematics teachers.

—Javad H. Zadeh, PhD, Brewton-Parker College and
Dalton State College, Georgia

I've been teaching math for nineteen years and NEVER seen anyone teach GCF or LCM the way you do in *Math Doesn't Suck*. I can't wait to share it with my students—I have recommended it to all of the teachers in the building! I also plan on passing my copy on to my 12-year-old daughter. Great job!

—Anne Park, Math Specialist,
Redding Middle School, Maryland

Just weeks ago I was complaining to a coworker how it's so easy to find elementary and high school "help" books, but how we can never find anything for middle school. A friend asked me if I'd heard of your book, and I bought it that same day. I teach a combination of special ed, ESL, "regular ed," and students that should probably be in honors math . . . ALL IN THE SAME CLASS! This book is perfect. Now I'm teaching these kids ways to do things that keep them ALL engaged and having fun. Thank you!

—Rebecca Choate, 7th Grade Math Teacher,
San Jacinto Junior High, Texas

I'm a math teacher at a very rural high school where math is definitely a four-letter word. I was always good at left-brained activities growing up—I never had trouble with math. I think because of this, I sometimes have trouble finding ways to vary instruction for my students. I picked up *Math Doesn't Suck* last weekend. My Pre-Algebra class was starting prime factorization and GCF on Monday, and because I had just finished reading those sections in your book, I had a couple more tools I could use. I am forever grateful for your insight into the young girls' math-o-phobia.

—Cheryl Ginnings, Math Teacher,
Golden City High School, MO

For alignment with NCTM and State standards,
visit mathdoesntsuck.com/standards.

MATH DOESN'T SUCK

How to Survive
Middle School Math
Without Losing Your Mind
or Breaking a Nail

DANICA McKELLAR

A PLUME BOOK

PLUME
Published by the Penguin Group
Penguin Group (USA) Inc., 375 Hudson Street, New York, New York 10014, U.S.A. •
Penguin Group (Canada), 90 Eglinton Avenue East, Suite 700, Toronto, Ontario, Canada
M4P 2Y3 (a division of Pearson Penguin Canada Inc.) • Penguin Books Ltd., 80 Strand,
London WC2R 0RL, England • Penguin Ireland, 25 St. Stephen's Green, Dublin 2, Ireland
(a division of Penguin Books Ltd.) • Penguin Group (Australia), 250 Camberwell Road,
Camberwell, Victoria 3124, Australia (a division of Pearson Australia Group Pty. Ltd.) •
Penguin Books India Pvt. Ltd., 11 Community Centre, Panchsheel Park, New Delhi –
110 017, India • Penguin Group (NZ), 67 Apollo Drive, Rosedale, North Shore 0632,
New Zealand (a division of Pearson New Zealand Ltd.) • Penguin Books (South Africa)
(Pty.) Ltd., 24 Sturdee Avenue, Rosebank, Johannesburg 2196, South Africa

Penguin Books Ltd., Registered Offices: 80 Strand, London WC2R 0RL, England

Published by Plume, a member of Penguin Group (USA) Inc.
Previously published in a Hudson Street Press edition.

First Plume Printing, July 2008
10 9 8 7 6 5 4 3 2 1

 REGISTERED TRADEMARK—MARCA REGISTRADA

The Library of Congress has catalogued the Hudson Street Press edition as follows:
McKellar, Danica.
  Math doesn't suck : how to survive middle school math without losing your mind or
breaking a nail / Danica McKellar.
    p. cm.
  ISBN 978-1-59463-039-2 (hc.)
  ISBN 978-0-452-28949-9 (pbk.)
  1. Mathematics—Study and teaching (Middle school)—United States. I. Title.
  QA13.M319 2007
  510—dc22

                        2007017091

Printed in the United States of America
Original hardcover design by Sabrina Bowers

Acknowledgments

$\mathcal{H}$ere are the people I want to thank, because life is a team sport!

Thank you to my parents, Mahaila and Chris, for never saying that math sucks and always encouraging me to pursue whatever I loved, even if those things had nothing to do with each other. And thank you to the rest of my very large and extended family, including Grammy, Opa, Crystal, Christopher, Connor, and everyone else—you know who you are!

Thank you to my wonderful literary agent, Laura Nolan, for asking me to write this book, and to the entire gang at the Creative Culture for holding my hand through my first experience as a book author. Thank you to Clare Ferraro, Laureen Rowland, my lovely and talented editor Danielle Friedman, and everyone else at Hudson Street Press and Penguin Books for wanting to put this book on the shelves. Thanks also to Sabrina Bowers, Abigail Powers, Susan Schwartz, Matthew Boezi, and Melissa Jacoby for their amazing work on the production and design end. Thank you to all the folks at NK

Graphics, who worked overtime to get this book to press in time, including Mary Page, Marcia Olander, Wendy Slate, Diane Luopa, Michele Jones, Rob Goetting, Jennifer Razee, and Robin Hogan. And a special thank-you to my new editor in chief, Luke Dempsey, for running with the relay stick with so much enthusiasm . . . and a really cool British accent.

Thank you to Hope Diamond for helping to shape my career (and life!) for the last nine years, and to everyone at Shepley Winings PR, including Bonnie Winings, Danielle Dusky, and Brenda Kelly, and also to the wonderful Math-A-Thon program at St. Jude Children's Research Hospital, and to Kenneth Chang and all the other journalists who have helped me find ways of sharing my love of math with the world.

Thank you to my lawyer, Jeff Bernstein, and also to Marc Wolf and to my manager, Adam Lewis, and Abrams Artists Agency and to everyone at CESD, including Cathey Lizzio and Pat Brady, for all being so understanding and supportive through every stage of my multifaceted interests over so many years!

Thank you to Kim Stern for being an overall incredibly supportive friend both in good times and those of crisis (like last-minute proofreading!), and to Crystal, Brandy, Todd, Damon, Andrew, Allen, Yvette, Dina, Ryan, Jen, Doc, Stephanie, and especially my mom and Mike Scafati for your incredibly helpful comments, and to my twelve-year-old goddaughter, Tori, for letting me know whenever a chapter needed a little "spicing up." Thank you to the incomparable Gocha & Shorena for teaching me to dance and saving me from getting hunched shoulders from sitting at the computer while working on this book!

Thank you to all the many, many middle school math teachers who let me interview them so I could better understand the "real deal" inside the classroom these days, helping me to shape this book and determine which subjects I would cover. Thank you to Shirley Stoll and everyone else who helped me gather such wonderful quotes from teenagers!

Finally, thank you to all of my math teachers from Westlake to UCLA, including Mrs. Jacobson, Prof. Varadarjan, the late Mr. Metzger, and of course my theorem collaborators at UCLA, Brandy Winn and Prof. Lincoln Chayes.

Lincoln, you brought the beauty and fun of mathematical thinking to life and challenged my mind in ways I'd never imagined possible. I'll always be grateful for you.

. . . and lastly, *thank you* to my incredibly loving and supportive sweetheart, Michael.

What's Inside?

PART VI: EVEN ALGEBRA DOESN'T SUCK

Math Used to Totally Suck

$\mathscr{I}$ was terrified of math.

I remember sitting in my seventh grade math class, staring at a quiz as if it were written in Chinese—it might as well have been a blank sheet of paper. Total brain freeze.

Nothing made sense, I felt sick to my stomach, and I could feel the blood draining from my face. I had studied so hard, but it didn't seem to make any difference—I barely even recognized the math problems on the page.

When the bell rang and my quiz was still blank, I wanted to disappear into my chair. I just didn't want to *exist*.

If you had told me that ten years later I would be graduating from college with a degree in mathematics, I would probably have told you to get your head examined.

As it turns out, though, no head examination necessary! I did in fact develop a love of math through the eighth grade and into high school, and made up tons of cool tricks and ways of remembering things along the way—tricks that I'm now going to share with you in this book!

In the pages that follow, you'll hear my adventures as a terrified math student, a confident actress, and everything in between. Best of all, you'll see how sharpening your brain will put you on the fast track to feeling fabulous in *all* areas of your life.

Oh yeah—I'll help you ace your next math test, too.

But Math Doesn't Suck!

Let's get a few things straight: Acne sucks. Mean people suck. Finding out that your boyfriend kissed another girl? That would totally suck. Too much homework, broken promises, detention, divorce, insecurities: suck, suck, suck, suck, suck.

But *math* is actually a good thing. Here are a few reasons why: Math builds confidence, keeps you from getting ripped off, makes you better at adjusting cookie recipes, understanding sports scores, budgeting and planning parties and vacations, interpreting how good a sale really is, and spending your allowance. It makes you feel smart when you walk into a room, prepares you for better-paying jobs, and helps you to think more logically.

Most of all, working on math sharpens your brain, actually *making you smarter* in all areas. Intelligence is real, it's lasting, and no one can take it away from you. Ever.

And take it from me, nothing can take the place of the confidence that comes from developing your intelligence—not beauty, or fame, or anything else "superficial."

When I was in middle school, I had insecurities like everybody else. It didn't help that I was on a TV series (*The Wonder Years*) at the time. Don't get me wrong—I loved acting, but it didn't take long for me to learn that when you are acting in front of millions of people you get a *lot* of attention that doesn't necessarily have anything to do with who you really are. Every day, walking down the street, people would come up to me, ask for my autograph, and tell me how much they loved the character I was playing. Great, right?

Well, after a few years of this, I started to wonder if people would still like me if I *weren't* on television. Eventually, whenever someone would tell me how much they liked my character, I would say "thank you," and then feel kind of empty inside. I started to question my self-worth.

I had a friend in high school who had beautiful, long, naturally red hair, and for years, everywhere she went, everyone told her how much they loved her long, red hair: friends, family, strangers, everyone. Finally, one day, when she was about seventeen years old, she showed up to school with her hair cut short—and dyed jet black!

She said she was tired of people complimenting her hair, and she needed to know what people liked about *her*. She had that same empty feeling on the inside that I did when people talked to me about being on TV. She wanted to be valued for something real, for what was on the inside. Of course, she was smart and funny and interesting—she just needed to figure that out for herself. And don't worry, her red hair did eventually grow back!

The good news is that the things that really matter, like our intelligence and personality—the things that feel good to be valued for— are things we have the ability to improve *ourselves*. While it's fun to focus on being fashionable and glamorous, it's also important to develop your smart and savvy side.

One of the best ways to sharpen your brain, and develop intelligence, is to study mathematics. It challenges and strengthens your mind in a way that very few other things do. It's like going to the gym—but for your brain!

I even took a break from acting for four years to go to college and major in mathematics, and it was one of the best choices I've ever made. These days, I've returned to acting, but with a new sense of confidence that came from developing my intelligence.

FAQs: How to Use This Book

What Kinds of Math Will this Book Teach Me?

This book focuses on the middle school math concepts that cause confusion year after year—fractions, rates, ratios, percents, etc. If you don't totally understand these concepts *now*, they will keep coming back to haunt you throughout high school—and beyond. So I thought I'd clear them up once and for all!

You'll notice some handwritten problems in the book . . . yes, this is my very own handwriting. Granted, my fourth grade teacher Mr. Williams never thought my penmanship was very good, but sometimes seeing stuff in actual handwriting makes it easier to understand, don't you think? (And I think Mr. Williams would agree that I've improved since then!)

You'll also notice that I don't include very many practice problems at the end of each section, mostly because I want to be able to give you as many tips and tricks as I can—and I figure that you are probably getting more than enough practice problems through your math class at school, right? But every single problem in this book has an answer—given in the back—and you can find detailed explanations on my website, mathdoesntsuck.com. That way you don't have to think to yourself, "But how did she *get* that answer?" Don't you just hate that? I do.

Do I Need to Read the Book from Beginning to End?

Nope! In fact, there are a few different ways to use this book:

- You can skip directly to the chapters that will help you with tonight's homework assignment or tomorrow's test.

- You can skip to the math concepts that have always been problem areas, to clear them up for good.

- Or you can, in fact, read this book from beginning to end and refer back to each chapter's "Takeaway Tips" for quick refreshers as you need them for assignments.

This Book Seems to Include More than Just Math—What's Up?

In addition to the math I teach, I've included a ton of extra features, including:

- "Real Life Testimonials" from gals just like you and me—hear their success stories!

- Math personality quizzes on pages 33, 110, and 235 to find out if you have "mathophobia," what your learning style is, whether you have trouble focusing—and what to do about it!

- A math horoscope on page 157, to find out what your sign can tell you about how *you* approach math, and how to deal!

- Quotes throughout the book from real kids and TV stars!

What Should I Already Know in Order to Understand This Book?

To get the most out of this book, it's a good idea to already be comfortable with both **times tables** and **long division**.

If you're not, no worries. On my website, mathdoesntsuck.com, you'll find a quick review of long division; and in the back of the book, you'll find multiplication tables. We all forget our times tables sometimes. For some reason, I could never remember what 7×8 was. One time I was doing a test, and I needed to know 7×8 for some fraction problem, and I just couldn't remember if it was 54 or 56 or 57! So I actually had to add them all up in a side column. That's right—I wrote down seven 8s and added them up. Not a pretty sight, but I was desperate.

7×8 is 56, by the way.

What If I'm Still Confused?

At the back of the book, you'll also find a "Troubleshooting Guide," which offers extra help with all of your math, well, troubles. Do any of these problems sound familiar?

- "Math bores me to death."

- "When it's time to do math, I get scared and try to avoid it."

- "I get confused and lost during class."

- "I think I understand something, but then I get the wrong answer in my homework."

- "My homework is fine, but when it comes time for a test, I freeze up and can't remember anything."

If so, flip to page 265 at any time, for solutions to all the above.

Let's get started!

Chapter 1

How to Make a Killing on eBay

Prime Numbers and Prime Factorization

*H*ave you ever made a friendship bracelet? I used to make them all the time. I loved going to the bead store and picking out pretty beads to string together. I haven't made one in awhile, but I have a friend who now makes a lot of money by making her own bracelets and selling them on eBay!

Let's make one: With medium-sized beads, you can usually make a whole bracelet using about 24 beads. Let's say you have 16 onyx beads and 8 jade beads. This bracelet is going to be beautiful!

Next we get to figure out what pattern we want to use. Well, let's divide up the beads into even groups, so we can see what our options are.

We could separate the 8 jade beads into

2 groups of 4 beads each,

OOOO OOOO

4 groups of 2 beads each,

◖◗ ◖◗ ◖◗ ◖◗

or 8 groups of 1 bead each.

◖ ◖ ◖ ◖ ◖ ◖ ◖ ◖

For the 16 onyx beads, we could separate them into

2 groups of 8 beads each,

●●●●●●●● ●●●●●●●●

4 groups of 4 beads each,

●●●● ●●●● ●●●● ●●●●

8 groups of 2 beads each,

●● ●● ●● ●● ●● ●● ●● ●●

or 16 groups of 1 bead each.

●●●●●●●●●●●●●●●●

There are only so many options for having even groups—and if we count the number of beads in each group, this actually tells us what the **factors** are.

What's It Called?

Factor

A *factor* of a number is a whole number that divides into the number evenly (with no remainders). For example, the factors of 16 are 1, 2, 4, 8, and 16. The factors of 3 are 1 and 3. The number itself and 1 are always factors of a number.

Based on the groupings we've made, look at some of the different patterns we could use to design the bracelet, if we use all 24 beads.

Notice that we can't have 5 even groups of beads if we want to use all 16 onyx beads. (Just try it yourself!) There would always be a group with not enough or too many beads. That's because 5 doesn't divide into 16 evenly; in other words, 5 is not a *factor* of 16.

"*M*ath keeps your brain energized. And smart people can do a lot more with their lives than people who don't exercise their brain. Period." Geena, 12

"*S*mart girls know themselves and take care of themselves. They have values, and morals, and they stick to them. They think before they act and are always trying to learn more. I admire smart girls." Marimar, 18

You've probably already learned about factors and prime numbers in school, but I'll do a little review here, because the ideas behind them will be very useful for the stuff we'll do in this book.

Prime Numbers . . . and Monkeys

Some numbers of beads cannot be *evenly divided up* no matter what you do. Lots of small numbers are like that: 2, 3, 5, 7. The only factors they have are 1 and themselves. There are bigger numbers like that, too, like 53 and 101. It's hard to believe that there's no way to evenly divide up 101, but it's true!

I like to think of these numbers as less "evolved" than most numbers. They don't have a whole lot going on upstairs, if you know what I mean. They're uncomplicated. They're "primitive," like monkeys. (Monkeys are a type of primate.) And perhaps that is why these less-evolved numbers are called **prime**.

Yes, prime numbers are a bit like monkeys. Just go with me on this, okay?

What's It Called?

Prime Number

A *prime number* is a number that has no factors except 1 and itself. In other words, no other whole numbers divide evenly into it. The first several prime numbers are: 2, 3, 5, 7, 11, 13, 17, 19, 23, and 29. There are many more, though!*

For technical reasons that are much more important to high-level math than to what this book covers, *1 is not considered a prime number*. Not a big deal—it's just a definition thing.

Prime Factor

You've found a *prime factor* of a number if it's prime *and* it divides evenly into the number. For example: 3 is a prime factor of 12 because 3 is a prime number *and* it is a factor of 12. On the other hand, 4 is a factor of 12, but because 4 is not a prime number (4 can be divided by 2), 4 is *not* a prime factor of 12.

Factoring

Factoring a number means finding its factors. Pretty simple, huh? What it really comes down to is "figuring out what you could divide a number by." For example, the factors of 6 are 1, 2, 3, and 6, because those are the numbers you can divide evenly into 6.

---

* *Did you know there's no biggest prime number? Yes, that means if you think you know what the biggest prime number is, no matter how big it is, I'll always be able to come up with a bigger one. For more on prime numbers, Google "prime numbers," "kids," and "math."*

How Many Lipsticks Does One Celebrity Need?

The gift bags given to guests at some Hollywood galas are extravagant. They often have multiple, top-of-the-line makeup kits inside a single bag!

Let's say you were in charge of assembling the gift bags for an event, and you had 18 extra lipsticks to divide however you'd like among some of the bags. You could distribute them—or *factor* them—like this:

$$18 = 9 \times 2$$

(2 gift bags with 9 lipsticks each, or 9 gift bags with 2 lipsticks each)

Or like this:

$$18 = 6 \times 3$$

(3 gift bags with 6 lipsticks each, or 6 gift bags with 3 lipsticks each)

Of course, you could also put 1 lipstick in 18 different bags or dump all 18 lipsticks into a single bag and take it home with you. That would be *factoring* 18 like this: $18 = 1 \times 18$.

As you saw in the bracelet examples, and now here, all factoring comes down to how we can divide stuff up *evenly*.

There are a few ways to factor a number. If you need *all* of the factors of a number, you can create a long list of all the numbers that divide into that number. For example, if you wanted to factor 16, you could write down: 1, 2, 4, 8, and 16 because those are all the numbers that divide evenly into 16. Or if you were factoring 18, you could write down: 1, 2, 3, 6, 9, and 18, because those are all the numbers that divide evenly into 18.

Factor Trees

In my opinion, the best tool for factoring numbers, especially if you want to find its **prime factors**, is something called a "factor tree."

Like monkeys swinging on the lowest branches of trees, prime numbers swing off the lowest branches of factor trees.* Let's say you want to factor the number 30.

Draw two little branches down from the number and ask yourself, "What are two numbers that when multiplied together give us 30?" Hey, how about 15 and 2?

Then you look at each of those new numbers and ask the same question. For 15, we can split it up into 3 and 5. But for 2, there's nothing else that divides into it except 1 and itself, which means it must be prime. Yes, like a monkey. We now circle the "monkeys" to keep track of them.

And what about 3 and 5? Each of them is a prime number, too, so they also get circled.

$$30 \rightarrow 30$$

And now we have factored 30 into its prime factors: 2, 3, and 5. Voilà!

QUICK NOTE! No matter how you *start* a factor tree for a particular number, you'll always end up with the same *prime factors* in the end. (If you've done it correctly!)

....

* *See? I told you to stick with me on this whole monkey thing. It's paying off now, isn't it?*

Let's factor the number 30 again, this time starting differently.

Again, begin by asking yourself, "What are two numbers that when multiplied together will give me 30?"

This time, however, let's answer, "Hey, how about 5 and 6!" Since 5 is prime, we can go ahead and circle it. What about 6? We can split it up into its two factors, 2 and 3.

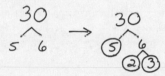

Each of the circled numbers is prime, so we're done! And we got the same prime factors of 30 as we did before: 2, 3, and 5. (It's sort of comforting to know that no matter how you start a factor tree, if you do it correctly it will always boil down to the same prime factors.)

Sometimes, you'll be asked to write a number as a *product* of its prime factors—this is called **prime factorization**.

What's It Called?

Prime Factorization

Prime factorization is the expression of a number as the product of its prime factors. For example, the prime factorization of 30 is $30 - 2 \times 3 \times 5$ (usually the prime factors are listed from smallest to biggest). It probably goes without saying, but in prime factorization all the numbers must be prime and they must multiply together to give you the original number.

QUICK NOTE! Notice that you can find bigger, nonprime factors of 30 by multiplying some of its prime factors together. The factors you get won't be prime factors, but they'll still be factors. For instance, if you multiply 2×3, you get 6, which is a factor of 30. Or you could multiply 3×5, and get 15, which is a factor of 30. You could also multiply 2×5 and get 10, which is another factor of 30. They're all numbers that divide evenly into 30, right? That means they are factors.

Doing the Math

Find the prime factorization of these numbers by doing a factor tree and then circling the monkeys—uh, I mean, the prime factors. Remember, there's more than one way to do most factor trees, so they might look different, but as long as you get the right prime factors, you're okay! I'll do the first one for you.

1. 72

<u>Working out the solution</u>: First I'll remember from my times tables that $8 \times 9 = 72$, so I'll start there.

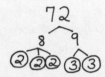

<u>Answer</u>: $72 = 2 \times 2 \times 2 \times 3 \times 3$. Done!

2. 15

3. 75

4. 100

5. 48

(Answers on p. 285)

You've Seen Him on TV!

"*I love smart girls. It always helps when a girl can hold up an intelligent conversation.*" Devon Werkheiser, Ned on Nickelodeon's *Ned's Declassified School Survival Guide*

EZ Divisibility Tricks for Factoring

With bigger numbers, it's often harder to see how to *start* the factor trees. For instance, if you have to factor the number 384, you know that you can start with 2, since it's even. But what about the number 567?

You could just start randomly dividing numbers into it and hope for the best, but that sounds time-consuming, and the last thing we want to do is spend unnecessary time doing math homework! Instead, here are some EZ divisibility tricks to help you detect right away if you can divide these factors into your number.

This chart can make factoring go *much* faster, once you're familiar with it. The best factor-divisibility tricks to memorize are the ones for 2, 3, and 5. They come up the most often. I'm also a big fan of 9.

Factor	Divisibility Trick
2	If the **last digit** of the number is even, then you know that 2 is a factor. *Example: 99,999,994 is divisible by 2, because the last digit is even.*
3	If the **sum of the digits** is divisible by 3, then you know 3 is a factor. *Example: 270. Add up the digits: $2 + 7 + 0 = 9$. Since 9 is divisible by 3, we know that 270 is divisible by 3.*
4	If the **last two digits** of the number are divisible by 4, then 4 is a factor. *Example: 712. Since 12 (the last two digits) is divisible by 4, we know that 712 is divisible by 4.*
5	If the **number ends** in 0 or 5, then you know 5 is a factor. *Example: 765 ends in a 5, so it's divisible by 5.*
6	If the **number passes the tests** for 2 and 3, then 6 is a factor. *Example: 504. It's even, and $5 + 0 + 4 = 9$, which is divisible by 3, so 504 is divisible by 2 and 3, which means 504 is divisible by 6.*
8	If the **last three digits** of the number are divisible by 8, then 8 is a factor. *Example: 70,008 is divisible by 8 because 008 is divisible by 8.*
9	If the **sum of the digits** is divisible by 9, then you know 9 is a factor. *Example: 981 is divisible by 9 because $9 + 8 + 1 = 18$, which is divisible by 9.*
10	If the **number ends in 0**, then you know that 10 is a factor. *Example: 111,110 ends in a 0, so 111,110 is divisible by 10.*

(There are tests for 7 and 11, too, but they're a little more compli-cated. See this book's website, mathdoesntsuck.com, if you want to check 'em out.)

Using this list of tricks, you might notice that 567 is actually di-visible by 9, because $5 + 6 + 7 = 18$, which is divisible by 9! In fact, $567 = 9 \times 63$. If you do the factor tree, you'll see that the prime fac-torization would be $567 = 3 \times 3 \times 3 \times 3 \times 7$. Not so bad, eh?

Doing the Math

Factor these big numbers using factor trees, and use the divisibility tricks chart to help! Be sure to circle the prime factors on the factor tree. Remember, there's more than one way to do a factor tree, but the circled prime factors should match the answer key. I'll do the first one for you.

1. 216

<u>Working out the solution</u>: I'm a big fan of adding up the digits to see if we get something divisible by 3 or, better yet, by 9. So, $2 + 1 + 6 = 9$. Yep! Let's divide by 9 and start the factor tree. And then it's pretty simple to fin-ish, since the 9 splits into 3s and 24 splits into 3 and 8, and 8 is just $2 \times 2 \times 2$.

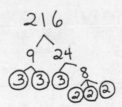

2. 105

3. 540

4. 1134

(Answers on p. 285)

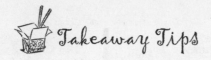
- A *prime number* is a number whose only factors are 1 and itself. Nothing else divides evenly into it. Rather "primitive" numbers, wouldn't you say?

- A factor is a whole number that divides evenly into another number; it's called a prime factor if it also happens to be prime. Just think of the "monkeys" swinging on the low branches of the factor trees—those are the prime factors, and you want to circle them. Monkeys like to be circled. They also like bananas, as I understand it.

- When you're not sure how to *start* factoring a number, just check the EZ divisibility tricks chart on p. 9 to make things go quickly!

Role Models

𝒯hink about who your role models are: Your mom, an older sister, a favorite aunt? Or maybe a teacher or your sports coach? Think of a woman in your life who just seems to have it all together—somebody you want to be like someday. There are probably one or two things that stand out the most about her: She's the best _____, or she always seems to succeed at _____. Maybe she always has a smile on her face, or she always seems to have the right thing to say, or she always works hard. Or maybe she just always seems calm and levelheaded.

These are the women you learn from and the women who will—just by their existence—help to shape you into the woman you'll become. Some of our role models are women we know personally, and some are public figures that we learn about from TV or the Internet.

One thing you'll find in common with all of your role models is *confidence*. Throughout this book you'll see real-life testimonials from fabulous women I admire. And what do they have in common with each other? That supreme confidence that comes from challenging yourself and feeling smart!

Is there someone in your life who inspires you to live up to your full potential of brains and beauty, inside and out? E-mail me about her at danica@mathdoesntsuck.com! Each month we'll post a few comments from readers on the website mathdoesntsuck.com, so check back often!

Chapter 2

Do You Still Have a Crush on Him?

Finding the Greatest Common Factor (GCF)*

*O*kay, so the last guy you had a crush on is history. You are totally over him. Now you have a whole new crush—and it feels great! He's much more your type. He's tall, funny, has brown hair, dimples . . . Come to think of it, he's a lot like the last guy. Wait a minute. Maybe you only like this guy because he reminds you of your old crush!

Darn it, maybe you haven't gotten over the last guy like you thought you had. Just how much *do* they have in common? Let's figure this out. If nothing else, it'll help determine what your "type" is. A lot factors into what makes someone attractive to you. Just what *are* those factors?

Old crush:
tall, dark hair, green eyes, dimples, funny, plays soccer, great smile

. . . .
* *Some textbooks call this the Greatest Common Divisor (GCD). They mean the same thing.*

New crush:

<u>tall</u>, <u>dark hair</u>, brown eyes, <u>dimples</u>, <u>funny</u>, plays piano, <u>great smile</u>

We've underlined the things they have in common, and now we can make a full list of the common factors that you are attracted to:

Common factors: tall, dark hair, dimples, funny, great smile

Now you try, with your *real* old crush and new crush. Write down their names, and list the traits that you like about each of them. Be honest—there are at least a few that they have in common!

Factors you liked about your old crush:

Factors you like about your new crush:

Now underline the things they have in common, and write those below.

List of your "type's" common factors:

What do you think? Do you have a "type"? And are you over the last guy? It's okay if you aren't, some crushes take awhile to get over—believe me, I've been there. I know your pain. I also know the pain of seemingly impossible math homework, and that kind of pain, believe it or not, is much easier to deal with.

Just like people, *numbers* can have factors in common, too. Let's try the same exercise again, only this time using 30 and 12 (instead of your old crush and new crush).

Keeping the factoring tips from chapter 1 in mind, begin by listing *all* their factors (not just the prime factors): that is, *all* the numbers that divide evenly into them. Then we'll underline the factors they have in common. Note that we don't have to include 1 as a factor. We can, but it's not necessary.

30: <u>2</u>, <u>3</u>, 5, <u>6</u>, 10, 15, 30

12: <u>2</u>, <u>3</u>, 4, <u>6</u>, 12

Common factors: 2, 3, 6

And the *greatest*—i.e., biggest—factor they have in common is 6. By the way, I've just demonstrated one method for finding the **greatest common factor** of two numbers. The greatest common factor (also known as the GCF) of 30 and 12 is 6.

In other words, 6 is the greatest (biggest) number that divides evenly into both 30 and 12.

What's It Called?

Greatest Common Factor (GCF)

The *greatest common factor* (GCF) of two numbers is the greatest (biggest) factor that they have in common. In other words, the GCF is the *biggest* number that divides evenly into *both* numbers.

QUICK NOTE! Sometimes the GCF of two numbers will be 1. (In other words, the biggest number that divides into both is 1.) When this is the case, they are then called **relatively prime**. Take 10 and 9, for example. Because the biggest factor they share is 1, they are considered relatively prime.

Knowing how to find the GCF of two numbers will come in handy in later chapters (especially when we start reducing fractions). In this chapter, I'm going to show you three ways to find the GCF. Each has its pros and cons, so when you're working on your math homework or a test, use whichever one *you* like the most! We'll start with the "crush" method we used above to compare old crushes and new crushes.

Method #1: The <u>G</u>reatest <u>C</u>rush <u>F</u>actor*

The Greatest Crush Factor method is great for finding the GCF of smaller numbers—but not as useful when the numbers get big, since it can be hard to think of *all* the factors of big numbers.

Step-By-Step

Finding the GCF of two numbers, using the Greatest Crush Factor method:

Step 1. List *all* the factors of each number. (But you don't have to include 1.)

Step 2. Underline the factors they have in common.

Step 3. The biggest, or "greatest," of those underlined factors is the GCF.

Watch Out! In the Greatest Crush Factor method, make sure you list *all* the factors of each number, not just the prime factors. For example, when listing the factors of 30, don't forget factors like 10 and 15. The biggest downfall of this method is that people forget to list *all* the factors of one of the numbers, listing only the prime factors instead—and then they get the wrong answer. Oops!

And...
Action! Step-By-Step in Action

Now let's do an example using this method: Let's find the GCF of 10 and 32. (Remember to write out *all* the factors—not just the prime ones!)

. . . .

* *I made up the name Greatest Crush Factor to help you remember GCF. You won't see that in your math textbook.*

Steps 1 and **2.**

10: <u>2</u>, 5, 10

32: <u>2</u>, 4, 8, 16, 32

Step 3. And now, of *all* the underlined common factors, the biggest is our GCF. It seems the only factor they have in common is 2, though—so the GCF of 10 and 32 is 2. Done!

 Doing the Math

Find the greatest common factor (GCF), using the Greatest Crush Factor method. I'll do the first one for you.

1. 36 and 20

<u>Working out the solution</u>: First, let's list all the factors of each number, and then underline the factors they have in common.

36: <u>2</u>, 3, <u>4</u>, 6, 9, 12, 18, 36

20: <u>2</u>, <u>4</u>, 5, 10, 20

The biggest number they have in common is 4.

<u>Answer</u>: the GCF of 36 and 20 is 4.

2. 70 and 14

3. 100 and 30

(Answers on p. 285)

What Do You Have to Say?

Method #2: The Multiplying Monkeys

I like to call this next method Multiplying Monkeys, because we use those swinging factor tree monkeys from chapter 1. (In that chapter, I explained how I like to think of prime factors as monkeys, because monkeys are primates. Prime-ates, get it?)

Step-By-Step

Finding the GCF of two numbers, using the Multiplying Monkeys method:

Step 1. Write out *factor trees* for both numbers, and *circle the prime factors*. (These are the monkeys swinging from the lowest branches.)

Step 2. Underline all the prime factors the two numbers have *in common.* If they have no prime factors in common, the GCF = 1.

Step 3. Make a list of the prime factors they have in common, including any repeats.

Step 4. *Multiply all* the common prime factors in this list, and you get the GCF! If they only have one prime factor in common, then *that* number is the GCF!

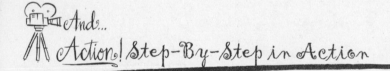
And...
Action! Step-By-Step in Action

Let's find the GCF of 20 and 24.

Step 1. Do a factor tree for each number, then circle the prime factors. (Remember, the prime factors are the monkeys at the very bottom of the branches, which are prime, and cannot be factored any more.)

Step 2. Now underline the prime factors the two numbers have in common. Let's see, they both have 2, *twice.* And . . . that's it.

Step 3. So we'll write down that the common prime factors are 2 and 2.

Step 4. Multiply 2 × 2, and we get our answer. The GCF of 20 and 24 is 4. In other words, the biggest number that evenly divides into both 20 and 24, is 4.

 °Watch Out! In the Monkey method, make sure that you include *all* the common prime factors in your list (step 3), even if there are repeats. For example, if both factor trees each have four 3s, then make sure you include all four 3s when you multiply the common prime factors to get the GCF. But make sure not to *double* the number of factors by mistake, either. For example, if each factor tree only has one 5, even though you're staring at a total of two circled 5s on your page, the number 5 should only go into your list of prime factors *once*. This is why I include step 3 in the step-by-step—so that you write down the prime factors that both trees have in common and don't get confused!

Doing the Math

Find the greatest common factor (GCF), using the Multiplying Monkeys method, following the step-by-step method on p. 18. I'll do the first one for you.

1. What is the GCF of 52 and 200?

<u>Working out the solution.</u> First let's do the factor trees. Hmm. I don't recognize 52 from my times tables, but I know that it's even, so I could start by dividing by 2. But I might also notice that 52 is divisible by 4, which when we divide into 52, gives us 13, which is prime. Next, we can split 200 into 10 and 20, and go from there.

As always, we circle the prime factors on the tree and underline the prime factors they have in common. Looks like the only prime factors they have in common are 2 and 2. So we multiply 2 × 2 = 4.

Answer: the GCF of 52 and 200 is 4.

2. What is the GCF of 90 and 135?

3. What is the GCF of 200 and 75?

(Answers on p. 286)

Method #3: The Birthday Cake

⟨ *Shortcut Alert!*

The Birthday Cake method is my personal favorite way to find the GCF of a number, because it's fast and works well no matter how big the numbers are. I think it's called the Birthday Cake method because when you're done you've drawn what looks sort of like an upside-down layered birthday cake. You know, a birthday cake where the layers and frosting are made of numbers? (Hey, I didn't name this one, okay?)*

Like most math shortcuts, this method is faster—and it also feels a little bit like magic when we get our answer. This is part of why we do the "Crush" and "Monkey" methods—because it's important to have a clear understanding of what the GCF of two numbers actually *is*, to solve other types of math problems later on. Once you understand that the GCF *is the biggest factor that divides evenly into two numbers*, then you can have your cake and eat it, too!

Let me show you what the Birthday Cake method looks like. Say we want to find the GCF of 24 and 18. Start by writing the two numbers side-by-side, and draw a little shelf or "layer" under them, starting on their upper left side. Then pick something that divides into

....

* *Yeah, yeah, I know. "Multiplying Monkeys Method" isn't exactly Shakespeare, either.*

both numbers. Let's say, 2, and put that on the side, divide each number by 2, and put the answers on the next "layer," underneath each number. When we divide 2 into 24 and 18, we get 12 and 9. Now draw a new layer and repeat the process. What goes into both 12 and 9? How about 3?

$$\underline{24 \quad 18} \;\rightarrow\; 2\underline{)24 \quad 18} \atop \;12 \quad\; 9 \;\rightarrow\; {2 \choose 3}\overline{)24 \quad 18 \atop \underline{12 \quad\; 9} \atop 4 \quad\; 3}$$

And now there's nothing left that we can divide into both 3 and 4, the newest "cake layer," so we stop.

Guess what the GCF is? Multiply the numbers along the side, and you've got it: $2 \times 3 = 6$. So the GCF of 24 and 18 is 6.

QUICK NOTE! The numbers you divide into the layers of the cake don't necessarily need to be *prime factors*. They can be any ol' factors you find that divide into both numbers! (For example, in the problem above we could have started the first layer with the number 6.) This is part of what makes the Birthday Cake method simpler than the others.

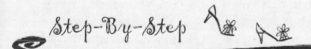

Step-By-Step

Finding the GCF of two numbers, using the Birthday Cake method:

Step 1. Start by writing the two numbers side-by-side.

Step 2. Draw a little shelf or "layer" under the numbers (see pictures above), and then pick a number that will divide into both numbers and write it on the left side of the layer. Then put the division answers directly below the numbers.

Step 3. Repeat step 2 with the new layer, until no more numbers will divide into the "newest" cake layer.

Step 4. Guess what the GCF is? Multiply the numbers along the left-hand side, and you've got it!

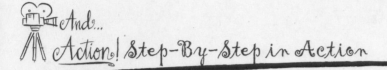

And...
Action! Step-By-Step in Action

What's the GCF of 72 and 180?

Step 1. We write the numbers side-by-side.

Step 2. Build the cake layer. Hmm, what number will divide into both 72 and 180? They are both even, so we could start with 2, but then I might notice that since $1 + 8 + 0 = 9$, which is divisible by 9, we know that 180 is divisible by 9 (see divisibility tricks on p. 9). And from our multiplication tables, we know that $8 \times 9 = 72$. So let's divide 9 into the cake!

Step 3. Keep going! Next, I might notice that both 8 and 20 are divisible by 4, so let's use 4 next.

$$
\begin{array}{c|cc}
9 & 72 & 180 \\
\hline
& 8 & 20
\end{array}
\quad \rightarrow \quad
\begin{array}{c|cc}
9 & 72 & 180 \\
4 & 8 & 20 \\
\hline
& 2 & 5
\end{array}
$$

We're left with 2 and 5 as our newest "cake layer," and since nothing divides into both of those numbers, we stop.

Step 4. Multiply the numbers along the side to get the GCF: $9 \times 4 = 36$.

(You also could have started with 2 and just ended up with more layers of cake. But the GCF would still be the same. Don't believe me? Give it a try!)

QUICK NOTE! If you need to find the GCF of three (or more) numbers, the birthday cake method still works great!

Doing the Math

Find the GCF of the following numbers, using the Birthday Cake method. I'll do the first one for you.

1. 104 and 78

<u>Working out the solution</u>: At first glance, I can tell they are both even, so let's start with 2. Dividing 2 into 104 and 78, we get 52 and 39. Next, since $3 + 9 = 12$, which is divisible by 3, then we know that 39 is divisible by 3. But 52 isn't divisible by 3, because $5 + 2 = 7$, which isn't divisible by 3 (remember, that divisibility trick only works for 3 and 9). Let's keep investigating. What other factors might 52 and 39 share? What happens when we divide 3 into 39? We get 13, which is prime. Well, does 13 divide into 52? If you try it, you'll see that 13 *does* divide into 52 evenly; in fact, $4 \times 13 = 52.$*

$$2\overline{\smash)\begin{array}{ll}104 & 78\end{array}} \qquad \rightarrow \qquad \begin{array}{l}\boxed{2}\overline{\smash)\begin{array}{ll}104 & 78\end{array}}\\ \boxed{13}\overline{\smash)\begin{array}{ll}52 & 39\end{array}}\\ \begin{array}{ll}4 & 3\end{array}\end{array}$$

So, continuing, we get 4 and 3, and nothing will divide into both of those, so we can stop. Multiplying along the left-hand side, we get $2 \times 13 = 26$.

<u>Answer</u>: the GCF of 104 and 78 is 26!

2. 80 and 104

3. 48 and 51 (Hint: add up each of their digits first, to check for a divisibility trick!)

4. 54, 180, and 90

(Answers on p. 286)

....

* *If you've ever played cards, you might have noticed that a deck of 52 cards has 4 suits and 13 different cards per suit, so every blackjack dealer in Vegas knows that $4 \times 13 = 52$. And now, so do you!*

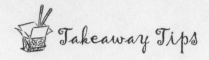 Takeaway Tips

- The *greatest common factor*, or GCF, of two numbers is the greatest (biggest) number that divides evenly into *both* numbers.

- In the Greatest Crush Factor method for finding the GCF, you list *all* the factors of *both* numbers, and then pick out the biggest one they have in common.

- In the Multiplying Monkeys method, you first create factor trees, then make a list of the *prime factors* both numbers have in common (including any common repeats), then multiply the prime factors together to get the GCF.

- The Birthday Cake method for finding the GCF involves creating "layers" beneath the two (or more) numbers; dividing various factors into the two numbers; then, eventually, multiplying the numbers along the left-hand side. This method is usually the quickest, and it's my personal favorite.

TESTIMONIAL:

Jen Stern (Los Angeles, CA)

<u>Before</u>: Terrified math student!
<u>Today</u>: Fourth grade teacher

Growing up, I hated math! I was one of
those kids who just didn't "get it." I had
tutors throughout most of high school.
Thankfully, all of them were very patient with me,
and they didn't make me feel like a complete idiot
for not understanding everything.

A specific math memory I have is being tutored by
my eleventh grade math teacher in Trig. She was so
kind and understanding. Even when she'd gone over
something with me a million times, she
acted as if it were the first time she
was explaining it.

Finally, during one of our tutoring
sessions, it was like a lightbulb went
off—everything just fell into place.
I was so excited—I got it! On my next
math test, I got a perfect score. I was
so proud of myself, and I could tell
that my teacher was, too.

> "I was one of those kids who just didn't 'get it.'"

Today, I'm a fourth grade teacher,
and I love it. I use math just about
every day. Not only do I teach the fourth grade math
curriculum, but I often work math into the other
subjects I teach, as well. For example, when my class
is reading a story for English, and it mentions
someone's date of birth, I'll ask my students to
figure out how old that person would be today—little
things like that, which I can throw in throughout the
day.

I guess the lesson here is: Don't give up, keep at
it, and eventually, you will "get it," too!

Chapter 3

You Can Never Have Too Many Shoes

Multiples and the Lowest Common Multiple (LCM)*

M̶y sister, Crystal, is a high-powered lawyer in New York City, and she's got a great sense of style. Her all-time *favorite* fashion item is shoes, and one of her favorite brands of shoes is Via Spiga.

Last summer, a few weeks before Crystal's birthday, I saw a pair of black and cream Via Spiga shoes online that seemed to shout, "Crystal!"—so I promptly ordered them. When she opened them later that month, she burst out laughing. It turns out that she had just bought the same shoes for herself the week before!

I assumed she would be returning them, but to my delight she beamed and said, "Oh, no. I'm not going to return these. I love them so much that I want **multiples** of them. That way, when one pair wears out, I'll have another pair ready to wear."

. . . .

* *Some textbooks call this the* Least *Common Multiple. They are the same thing.*

I was thrilled! And then I realized that I have multiples of a few items in my closet, too. I've got 4 of the same V-neck shirt (in different colors), 5 pairs of my favorite black tights, and 3 of the same gray tank top. Maybe I'll take a cue from Crystal and, the next time I find the perfect pair of black heels, I'll buy multiples. (They do go with everything, from jeans to the "little black dress.")

This got me thinking: When I find these perfect black heels, how many pairs should I buy? My closet space is limited, so I'll have to think about how many shoes I can squeeze in there. If I buy two pairs, then I'll have a total of 4 shoes, right? And if I buy three pairs, I'll have 6 shoes. If someone found the "perfect" black heel and went on a crazy shopping spree, they might end up with any of these multiples of 2: 2, 4, 6, 8, 10, 12, 14, and so on. Can you even imagine having 14 of the same shoe in your closet, looking back at you? Well, if you had 7 pairs of the same shoe, that's what you'd see!

In life, *multiple* means "more than one of a particular thing." In math, *multiple* means "more than one of a particular *number*."

What's It Called?

Multiple

A *multiple* is the product of two counting numbers. For example, some multiples of 2 are 2, 4, 6, 8, and 10.

Here are some of the multiples of 3 (which happens to be my favorite number).

3, 6, 9, 12, 15

We got this list by doing:

$3 \times 1 = 3, \quad 3 \times 2 = 6, \quad 3 \times 3 = 9, \quad 3 \times 4 = 12, \quad 3 \times 5 = 15$

and so on. See? To find the *multiples* of 3 above, we just *multiplied* different numbers times 3. In the same way, we can find multiples of 8 by multiplying different numbers times 8. Some multiples of 8 are 8, 16, 24, 32, 40, 48, 56, 64, 72, 80, 88, 96, and 104.

Doing the Math

List the first 10 multiples of these numbers (and try not to look at the multiplication table in the back!). I'll do the first one for you.

1. 4

<u>Answer:</u> 4, 8, 12, 16, 20, 24, 28, 32, 36, 40

2. 5

3. 7

4. 12

(Answers on p. 286)

QUICK NOTE! Factors and multiples are opposites of each other. For example, since 4 is a factor of 20, we automatically know that 20 is a multiple of 4. Remember, multiples are the bigger ones—which makes sense, because to have multiples of anything means to have lots!

Lowest Common Multiple

If you list out the multiples of *two* numbers, then you may notice that they have some multiples *in common* with each other. For example, let's list some multiples of 6 and 9.

6: 6, 12, <u>18</u>, 24, 30, <u>36</u>, 42, 48, <u>54</u> . . .
9: 9, <u>18</u>, 27, <u>36</u>, 45, <u>54</u>, 63, 72, 81 . . .

Notice that so far, they already have 18, 36, and 54 in common. They would have many more if we kept going—but we can see that the *smallest* multiple they have in common is 18.

What's It Called?

Lowest Common Multiple (LCM)

The *lowest common multiple* is the <u>smallest</u> multiple that two numbers share in common.

QUICK NOTE! If two numbers have no factors in common, then the LCM of those two numbers will just be the product of those two numbers. For example:

4: 4, 8, 12, 16, <u>20</u>, 24, 28, 32, 36, <u>40</u>, 44 . . .
5: 5, 10, 15, <u>20</u>, 25, 30, 35, <u>40</u>, 45, 50 . . .

Since 4 and 5 have no common factors (in other words, they are *relatively prime*), their LCM is their product: 4 × 5 = 20. Just something helpful to keep in mind!

Finding LCMs will be very helpful for adding and subtracting fractions with different denominators . . . which we'll be doing a ton of in chapter 8. This calls for a shortcut, don't you think?

Shortcut Alert!

USING THE BIRTHDAY CAKE METHOD TO FIND THE LOWEST COMMON MULTIPLE

Remember the Birthday Cake method we used (on p. 20) to find the GCF (greatest common factor) of some numbers? We can also use the Birthday Cake method to find the *LCM* of two numbers!

Let's use the Birthday Cake method to factor the numbers 8 and 12. Now, if we wanted the good ol' GCF of 8 and 12, we would just look at the numbers on the *side* of the cake and multiply them together. Try it! That would give us $2 \times 2 = 4$. So the GCF is 4.

But how about the LCM? Notice how the outside numbers in the cake form the shape of an L.

$$2 \overline{\smash{\big)}\begin{array}{cc} 8 & 12 \\ 4 & 6 \end{array}} \quad \rightarrow \quad \begin{array}{c|cc} 2 & 8 & 12 \\ 2 & 4 & 6 \\ \hline & 2 & 3 \end{array}$$

Well, if you multiply *all* the numbers that make up the big L, you'll get $2 \times 2 \times 2 \times 3 = 24$. And the LCM of 8 and 12 is 24. Just remember the L for LCM!

You can also test this method by doing our original "listing all the multiples out" method.

8: 8, 16, <u>24</u>, 32, 40, <u>48</u> . . .
12: 12, <u>24</u>, 36, <u>48</u>, 60 . . .

And happily, we get the same LCM, 24!

Take Two! Another Example!

Now let's do the Birthday Cake method on the numbers 4 and 5.

Hmm. There's nothing to start with, because nothing divides into both 4 and 5. Well, we can still divide by 1, right? Let's try that:

$$\lfloor 4 \quad 5 \quad\quad \longrightarrow \quad 1 \lfloor 4 \quad 5 \atop \overline{4 \quad 5}$$

Okay, that didn't feel very productive. But you know what? Let's go ahead and multiply the numbers that make up the big L and see what we get.

$$1 \times 4 \times 5 = 20$$

And we know from our example (page 29) that 20 is indeed the LCM of 4 and 5. Voilà!

QUICK NOTE! You can use the cake method on three or more numbers to find their GCF, but for find-ing the LCM, you can only use the cake method for two numbers at a time! After all, if you had three numbers along the bottom, when you went to circle the big "L" for LCM, it wouldn't look much like an "L" anymore, with that overly long bottom part, would it?

Take Three! Yet Another Example!

Find the LCM of 4, 6, and 20, using the birthday cake method:

We can do only *two numbers* at a time for the LCM with the cake method. It doesn't matter *which* two numbers we start with, so let's do 4 and 6, and we get **12**. Then we take that **12** and we use the cake method again, this time to find the LCM of 12 and our leftover number, 20. And the answer will be the LCM of all three numbers!

$$2 \lfloor 4 \quad 6 \atop \overline{2 \quad 3} \qquad 4 \lfloor 12 \quad 20 \atop \overline{3 \quad 5} \qquad 4 \times 3 \times 5 = \boxed{60}$$
$$\hookrightarrow 2 \times 2 \times 3 = \underline{12}$$

Answer: The LCM of 4, 6, and 20 is **60**.

If we had tried to use the cake method on all three numbers at once, it would have been a disaster. Try it yourself! We'd end up multiplying too many numbers and get 120—yikes!

Doing the Math

Find the LCMs (lowest common multiples) of these numbers, first by listing out multiples and underlining the common ones, and then by doing the Birthday Cake method and multiplying along the big L. I'll do the first one.

1. 6 and 8

<u>Working out the solution</u>: First let's write out the first few multiples of each.

6: 6, 12, 18, <u>24</u>, 30

8: 8, 16, <u>24</u>, 32, 40

<u>Answer</u>: the LCM of 6 and 8 is 24.

Now let's do the birthday cake method.

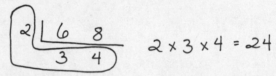

$2 \times 3 \times 4 = 24$

Since we multiply along the big L for the LCM, and $2 \times 3 \times 4 = 24$, the LCM is 24. As you can see, our two answers are the same. Phew!

2. 9 and 12

3. 6 and 7

4. 4 and 16

5. 9 and 15

(Answers on pp. 286–7)

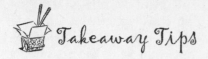 *Takeaway Tips*

- In math, *multiples* of a number are what you get when you multiply that number times counting numbers like 1, 2, 3, 4, 5 . . .

- *Factors* and *multiples* are opposites of each other—multiples are the bigger ones, of course!

- The *lowest common multiple* (LCM) of two numbers is the smallest number that is a multiple of both. Note that if the two numbers have no common factors, the LCM will be the product of the original two numbers multiplied together.

- I like using the birthday cake method for finding the LCM of two numbers. And remember, if you use the cake method for three numbers, you can do only *two numbers at a time!* (Or the circled numbers won't look like an "L" anymore!)

QUIZ #1: Are You a Mathophobe?

Does math tantalize or terrify you? Do you quiver at the mention of the "m" word? Let's see what expert psychologist Robyn Landow, PhD, says. Take her quiz and see how you fare!

1. You just left a math test and are chatting with a friend about how you think you did. You:

   a. Know you didn't totally bomb but are a little bummed out because, just as you feared, there were too many questions on that one topic you've been really confused about!

   b. Are confident. You studied, came in feeling as prepared as you could, and finished the test before the bell rang. As long as you didn't make a bunch of careless mistakes, you think you did all right.

   c. Can't even talk about it. *Can't people see on your face that you don't want to talk about it?*

2. You like being active in your school's student council—you're currently a class rep. One of your best girlfriends is president, and one afternoon she calls to tell you that this year's treasurer has to step down, and she is begging you to fill in for the rest of the year. The treasurer is in charge of doing all the accounting—in other words: math. You:

    a. Politely find a way to say "no way!" You understand her need, but you are just not the person for this job.

    b. Hesitate. Math is not your thing, but there are plenty of people to ask for advice if it gets tough, and you want to come through for the group. In the end, you say yes.

    c. Tell her you will do it—no problem!

3. You have a ton of homework tonight in all of your classes and aren't really sure where to start. You:

    a. Start with math. You don't want to be too tired when it comes time to memorize those formulas, and there is no time like the present. You might even have time at the end of the night to go over it again.

    b. Avoid math until the very end. Why start out making your night difficult? You're not even sure if you will know how to do it, anyway.

    c. Know that math is looming in your book bag but choose to go with another topic first. You know you will get to it, but maybe you should finish the easier topics first.

4. That cute boy next to you at the lunch table is giving you his phone number so you can call him the next time you go ice skating with a group of friends. But wait—you don't have a pen! You:

    a. Do your best to remember it, but also give him yours. Chances are one of you will remember the phone number!

    b. Recite it in your head or imagine yourself dialing the numbers to help yourself remember it, as you run over to your friend a few tables over to ask her for a pen so you can jot it down.

    c. Pretend to be memorizing it but realize that will never work. You know that your best friend's mom works with his dad, and you decide that you will figure out how to get his number that way.

5. It is first period math, and the teacher just announced a pop quiz! You got home late last night and barely had time to complete your homework, much less look over your notes. You:

    a. Take a deep breath and tell yourself that you will do the best you can. What else can you do?

    b. Tense up. You are already struggling with this topic in math, and a pop quiz is all you need! But after a few minutes, you close your eyes and try to remember everything you did in class yesterday. And after a quick scan of the quiz, you see that there are several questions you know how to do.

    c. Panic. Maybe you shouldn't even try. After all, you're not going to get anything right.

6. It is time for your quarterly meeting with your school advisor, and she asks you how you are doing in math this semester. You respond:

    a. With silence. You feel like you want to cry. This is the first time anyone has actually asked you about math—and you realize that you finally need to tell someone you are getting lost.

    b. "It's going fine. Sometimes I get confused, but I get extra help after school when I need it."

    c. "Why do you ask? Did the teacher say something to you? Did she say I've been looking a little confused these past few weeks?"

7. Your math teacher announces that this next topic is going to be "even more difficult" than the previous ones you have discussed. You think:

    a. "What? I bet this is going to be tough. I've gotten through the semester so far, though, and I will charge on."

    b. "Ugh! How on earth am I going to do this? Is it possible to ask for a transfer to another class that is not going to study this topic?"

    c. "Wow, this is going to be challenging, but I like a good challenge. I better make sure I take excellent notes and mark the areas I may need to review."

8. When it comes to tests in school, you:

    a. Feel nearly the same about them in all subjects. Nobody looks forward to tests, but they are a part of life.

    b. Find it easier to study/take tests that are in subjects other than math. There is just something about math that makes you feel uneasy.

    c. Hate math tests. You don't know how to study for them, and even when you do study, you tend to forget most of what you reviewed when you sit down to take the test.

9. You would describe your participation in math class as:

    a. Somewhat involved. Although you don't always speak up when you don't understand something, you will sometimes ask the teacher a question after class or call a friend that night for homework help.

    b. Active. You come to class prepared, and if you have a question, you ask it.

    c. Nada. You tend to zone out, and then you want to disappear when you are asked a question or need to go up to the board.

10. When it comes to math study groups with your friends, you:

    a. Avoid them like the plague. But then, you avoid anything having to do with math like the plague.

    b. Go, but they are not always helpful. In fact, at times you've found that studying math with a group can make things worse—like when everyone else is also confused, or decides to turn it into socializing instead. But how else are you supposed to get help?

    c. Go to them when you think they will be genuinely helpful. If you feel like you'll get more out of studying a concept by yourself, you know when to stay away. When it would be helpful to talk through a concept with a friend, though, you are not only there—you are hosting the group!

Scoring:

1. a. 2;	b. 3;	c. 1	**6.** a. 1;	b. 3;	c. 2
2. a. 1;	b. 2;	c. 3	**7.** a. 2;	b. 1;	c. 3
3. a. 3;	b. 1;	c. 2	**8.** a. 3;	b. 2;	c. 1
4. a. 2;	b. 3;	c. 1	**9.** a. 2;	b. 3;	c. 1
5. a. 3;	b. 2;	c. 1	**10.** a. 1;	b. 2;	c. 3

25–30 points

You are cool as a cucumber! Even if math is not your strongest subject, you are comfortable with it and know that rewards come to those who work hard. You do not shy away from situations where numbers are involved, and you are a role model to your friends. Keep up the excellent attitude!

16–24 points

Uh-oh—your fears about math may be getting in the way of your learning; it seems you have difficulty getting comfortable with the subject. When you feel nervous, stay positive. Instead of saying, "I am going to fail" while studying for a test, focus on how hard you are preparing for it—tell yourself, "success can be mine." Study with friends who demonstrate confidence and comfort with math—their attitude can be contagious!

10–15 points

Okay, this attitude about math has gotten out of control! Whether your troubles started out as simple worry about math or with some worse-than-expected grades on early tests, you have put yourself in a place where you just don't see yourself doing well. It's time to have a serious talk with your math teacher, to let him or her know what is going on, and to get some advice for how to get yourself back on track. He or she may be willing to spend some time helping you out with the tricky spots or recommend an after-school learning buddy or tutor. And let your parents in on this issue as well. Maybe they had similar feelings when they were your age—and now your dad is an engineer or your mom is an accountant. Also, congratulate yourself for even picking up this book. Says Danica, *You're off to a great start—keep this book by your side, and we'll get through this together!*

Chapter 4

Everything You Ever Wanted to Know About Pizza but Were Afraid to Ask

Introduction to Fractions and Mixed Numbers

Ah, the fraction. Throughout the course of history, these innocuous little numbers have conjured up all sorts of feelings: Hesitation, fear, dread . . . hunger.

Look, fractions may seem scary at first, but they're not so bad when you see them for what they really are—pizzas!

Here's what a **fraction** looks like.

$$\text{numerator} \rightarrow \frac{2}{3} \qquad \leftarrow \text{little fraction "dividing line"}$$
$$\text{denominator} \rightarrow$$

Usually the top and bottom are whole numbers (like 1, 2, 3, 4, 5, and so on). Fractions are used to describe *parts of a whole*. Your textbook probably tells you to think of $\frac{3}{8}$ as 3 "parts" out of 8 "parts total."

Personally, when I work on fractions, I prefer to think of pizza. But then, I really like pizza. (C'mon, who doesn't like pizza?)

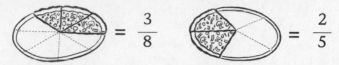

$$= \frac{3}{8} \qquad = \frac{2}{5}$$

So $\frac{3}{8}$ represents 3 slices of pizza out of 8 slices total, and $\frac{2}{5}$ represents 2 slices out of 5 total slices.

The **denominator** (bottom) is the total number of slices it would take to make up a *whole pizza*, and the **numerator** (top) is the number of actual *slices* of pizza that we have.

"ring ring"

What's It Called?

Numerator and Denominator

With all of the math vocabulary you're expected to remember, it's easy to confuse which part of the fraction is the *numerator* and which is the *denominator*. This will help:

The *denominator* is the bottom part of the fraction. So, in $\frac{2}{3}$, the 3 is the *denominator*.

It's *down below* the fraction line. Just remember the *d*: the *denominator* is *down* below. See? Helpful, right?

"*Math* problems can be hard, but at some point, I get it. Math isn't something that you don't understand forever." Vanessa, 12

"*I* used to hate math—it used to be my worst subject. Now it's one of my best." Tiffany, 16

"*Math* is everywhere. . . . It keeps your mind sharp and helps facilitate many other skills." Marimar, 18

What Do You Have to Say?

Doing the Math

Write down the fraction that best reflects the situation I describe. I'll do the first one for you.

1. The baseball game has 9 innings total, and we have watched 4 so far. What fraction of the game have we watched?

We've watched $\frac{4}{9}$ of the game

2. My best friend needs to borrow a dress from me, but she only likes to wear black. I have 5 dresses in my closet, but only 2 of them are black. What fraction of my dresses is my best friend going to be interested in?

3. While IM'ing, I notice that only 7 of my buddies are online, and I have a total of 73 buddies. What fraction of my buddies is online?

4. It rained 3 days last week and was sunny the rest of the time. What fraction of the days were sunny? (Read the question carefully!)

(Answers on p. 287)

More than One: Improper Fractions vs. Mixed Numbers

Sometimes we'll see a fraction with a bigger top (numerator) than bottom (denominator), like $\frac{7}{5}$, for instance. Strange as it may look, this is a fraction, too! And these types of fractions obey all the same rules as fractions with smaller numerators.

"Hold on a minute," you might be thinking. How can the top be bigger, when the bottom was the *whole* and the top was *part* of the whole? How can you have more parts than the whole?

It's simple. To use our pizza example from above: if you have $\frac{7}{5}$ pizzas, then that just means you have *more than one* pizza!

The fraction $\frac{7}{5}$ means that we have 7 slices of pizza. And, since we can see from the denominator that the pizzas are cut into 5 slices each, it means we have a whole pizza, *plus 2 extra slices*. In fact, you could rewrite that fraction as 1 pizza, plus $\frac{2}{5}$ of a pizza. Or you could just write them together as $1\frac{2}{5}$ (this is read out loud like, "One and two-fifths"). Note that $\frac{7}{5}$ and $1\frac{2}{5}$ have the same value. In other words, they are two different ways of describing the *same amount* of pizza.

What's It Called?

Mixed Numbers and Improper Fractions

A **mixed number** is a whole number and a fraction together. Some examples of mixed numbers include $1\frac{2}{11}$, $5\frac{3}{4}$, and $2\frac{7}{9}$. Mixed numbers are called "mixed" because you're *mixing* whole numbers with fractions. You would read $2\frac{7}{9}$ out loud as "two and seven ninths" and note that $2\frac{7}{9} = 2 + \frac{7}{9}$; they have the same value.

An **improper fraction** is a fraction with a numerator greater than or equal to its denominator. Some examples of improper fractions include $\frac{5}{2}$, $\frac{13}{7}$, and $\frac{100}{99}$. Improper fractions are called "improper" because they're top-heavy, and if they were to fall down, that would be highly improper, especially at a British tea party. Okay, I'm not really sure why they're called "improper," but I like imagining a top-heavy fraction waddling into a tea party, falling over onto the pastries, and some lady exclaiming (in a British accent), "How improper!" This image helps me to remember what these types of fractions are called, since, after all, improper fractions have bigger "tops" than "bottoms."

Going Back and Forth between Improper Fractions and Mixed Numbers

As you saw in the pizza example above, for every improper fraction there is a mixed number that has the same value—and vice versa. It's important to know how to go back and forth between the two—without having to use pizza. Can't you just imagine your teacher's face if you were to turn in a homework assignment done entirely in pizza drawings?

Improper Fractions → Mixed Numbers

If we have $\frac{7}{4}$ of a pizza, then we can tell that we have more than one pizza (because the numerator, 7, is bigger than the denominator, 4, right?). But just how many *whole* pizzas do we have, and how many leftover *slices* are remaining? In other words, can you write the improper fraction $\frac{7}{4}$ as a mixed number?

I'll let you in on a little secret: ***Fractions are division problems in disguise!*** Here's the trick: Treat $\frac{7}{4}$ like a *division problem*, and you can easily convert it into a mixed number. You can read from top to bottom: "7 divided by 4."

$$\frac{7}{4} = \text{divided by} \begin{array}{c} 7 \\ 4 \end{array} = 4\overline{)7}$$

Or here's another way to think of it: Improper fractions are top-heavy, right? They have a bigger top than bottom—so just *tip* that sucker over, and divide.

push
$\rightarrow \dfrac{7}{4} \rightarrow$ ⟋7 ⟍4 ⟍ Ah! I'm falling! $\longrightarrow 4\overline{)7}$

Dividing, we get: $4\overline{)7}^{\,1\ R3}$

With an answer of 1 R3, we get 1 whole pizza, plus 3 *remaining* slices. But instead of writing 3 as a "remainder," you can also write it as a fraction. You just put the remainder over the same denominator you started with. The R3 becomes $\frac{3}{4}$. So our answer is that we have 1 whole pizza, plus 3 leftover slices—in other words, $1\frac{3}{4}$ pizzas. Let's break it down into steps:

· · · · · · · · · · · · · · · · · · · ·
* *In fact, the symbol $\frac{7}{4}$ represents the value you get when you divide $7 \div 4$. This is true for all fractions: $\frac{1}{2} = 0.5$ because that's what you get when you divide $1 \div 2$.*

Step-By-Step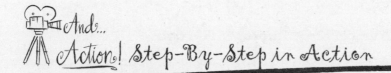

Converting improper fractions → mixed numbers:

Step 1. Tip the fraction over → and divide the denominator (bottom) into the numerator (top).

Step 2. The division answer will give you the *whole number* part of the mixed number, and a "remainder," if there is one. The remainder equals the number of *remaining* slices, so it becomes the numerator of the *little fraction* part of the mixed number. (The denominator of the little fraction should be the same as what you started with.)

 And...

Action! Step-By-Step in Action

Convert $\frac{18}{7}$ into a mixed number.

Step 1. Tip it over → and divide: $7\overline{)18}^{\;2\;R4}$ We get 2 whole "pizzas," plus 4 "remaining slices."

Step 2. Put the remainder over the same denominator you started with, and you get the answer $\frac{18}{7} = 2\frac{4}{7}$. Done!

 QUICK NOTE! When you are switching back and forth between improper fractions and mixed numbers, remember that you aren't changing the *value* of the number; you've just rewriting it in a different form.

Doing the Math

Rewrite these improper fractions as mixed numbers. I'll do the first one for you.

1. $\frac{9}{2} =$

Mixed Numbers ➡ Improper Fractions

Let's say you're having a bunch of friends over, and your mom ordered 5 pizzas. Each pizza is divided into 5 slices, like this:

How many slices total will you have for your guests? Well, 5 pizzas with 5 slices each—that's $5 \times 5 = 25$ slices total. Great!

But now let's say your Chihuahua, Sparky, got into the pizzas and your mom had to throw some of the slices away. She tells you that there are only $4\frac{2}{5}$ pizzas left. How many total slices will there be for your friends now?

In order to figure this out, we need to convert $4\frac{2}{5}$ into an improper fraction. Once we do, the top number (numerator) will tell us the *number* of slices we have left.

First, notice that $4\frac{2}{5}$ ("four and two-fifths") is the same thing as $4 + \frac{2}{5}$. This represents "4 pizzas, plus 2 slices of pizza." So how many slices are in the 4 pizzas?

Well, each pizza is made up of 5 slices, and we have 4 pizzas total, so that's $4 \times 5 = 20$ slices total. Next we add those remaining 2 slices (20 slices + 2 slices), which gives us a total of 22 slices of pizza.

$\underset{\text{20 slices}}{\left(4\right)} + \underset{\text{2 slices}}{\left(\tfrac{2}{5}\right)}$

In other words, we have $\frac{22}{5}$ pizzas. And now we have shown that $4\frac{2}{5} = \frac{22}{5}$.

Hmm, that made sense, but it took awhile. Here's a shortcut for converting mixed numbers to improper fractions, *without* having to draw pictures. I like to call it the MAD Face method. You'll see why in a moment.

Step-By-Step 👠 👠

Converting mixed numbers → improper fractions, using the MAD Face method:

Step 1. Multiply the denominator times the whole number.

Step 2. Add that to the numerator.

Step 3. Denominator stays the same. Done!

To me, that arc has always looked a little like a frowny face—like a *mad* face, with the "×" and the "+" as the eyes. Just tilt your head to the left. See it?

MAD:
Multiply
Add
Denominator stays the same

If you draw the arc and remember the MAD Face, you'll never be confused about how to go from mixed numbers to improper fractions again!

And...
Action! Step-By-Step in Action

Let's do the pizza problem above again, this time without *drawing pictures. Convert $4\frac{2}{5}$ to an improper fraction, using the MAD Face method.*

$$\overset{+}{\underset{\times}{\Big(}} \!\!\! \overset{\curvearrowright}{4} \; \frac{2}{5}$$

Step 1. <u>M</u>ultiply the denominator times the whole number: $5 \times 4 = 20$.

Step 2. <u>A</u>dd that to the numerator: $20 + 2 = 22$.

Step 3. Put the result on top of the <u>d</u>enominator, which stays the same as before, so we get $\frac{22}{5}$. And now we have shown that $4\frac{2}{5} = \frac{22}{5}$. Done!

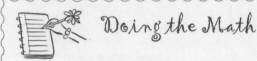 *Doing the Math*

Rewrite these mixed numbers as improper fractions. I'll do the first one for you.

1. $3\frac{4}{7} =$

<u>Working out the solution</u>:

$$+ \left(\mathrel{\underset{\times}{}} 3\frac{4}{7} \right.$$

M: <u>M</u>ultiply the denominator by the whole number, $7 \times 3 = 21$

A: <u>A</u>dd that to the numerator, $21 + 4 = 25$

D: Put it over the same <u>d</u>enominator, $\frac{25}{7}$

<u>Answer</u>: $3\frac{4}{7} = \frac{25}{7}$

2. $2\frac{1}{2} =$

3. $6\frac{2}{3} =$

4. $1\frac{3}{5} =$

(Answers on p. 287)

Why Improper Fractions Are Like Tennis Shoes

You may be wondering why you have to learn two ways of writing down these "big" fractions, whose values are more than 1. It's simple: mixed numbers and improper fractions serve different purposes. Each is good for different things, that's all.

Imagine you had only two pairs of shoes—a pair of cute tennis shoes and a pair of beautiful black-suede high heels. What if someone asked you, "Why do you need two pairs of shoes?" Realizing that you were talking to a complete moron, you'd politely explain, "Because I use them for different things. Most of the time I wear my tennis shoes, because I can wear them almost everywhere. But if I'm getting dressed up, I use my high heels." It's the same with improper fractions and mixed numbers.

Improper fractions = tennis shoes
Mixed numbers = high heels

Anytime you need to "do" something—add it, subtract it, multiply it, divide it, convert it to a decimal or percent, and so on—it's usually *much easier* to use improper fractions. (I strongly suggest changing mixed numbers into improper fractions before attempting to "do" anything with them.) For instance, if you have to multiply $1\frac{2}{5} \times 3\frac{2}{3}$, you should first convert the mixed numbers to improper fractions, and *then* multiply them ($\frac{7}{5} \times \frac{11}{3}$) because now you can just multiply across the top and bottom to get your answer (which is $\frac{77}{15}$). While you *can* use mixed numbers to solve addition and subtraction problems, they are usually easier to handle as improper fractions.

Like I said, improper fractions are sort of like the tennis shoes—they work almost everywhere, and they're much easier to use. (And tennis shoes, appropriately enough, would also be considered "improper" at a British tea party, wouldn't they?)

Mixed numbers, however, are more like high heels, because they are useful in only a few situations. However, mixed numbers are more "presentational." They are much better for communicating ideas in real life. After all, if you wanted a sandwich and a half for lunch, you wouldn't scribble a note to your mom on the refrigerator asking if you could please have "$\frac{3}{2}$ turkey sandwiches for lunch." Instead, you'd ask for "$1\frac{1}{2}$ turkey sandwiches" (a mixed number).

Mixed numbers are better for your *final answer on word problems* because they more clearly describe the answer to the question. Here's an example: Maggie drove $\frac{3}{5}$ of a mile to the grocery store then drove back home, another $\frac{3}{5}$ miles. How many miles did Maggie drive total? So, you'd add the two distances she drove together, and get $\frac{3}{5} + \frac{3}{5} = \frac{6}{5}$. Your answer could be, "Maggie drove $\frac{6}{5}$ of a mile." But you can imagine the blank stares Maggie would receive if, when she got home, she told her sister, "Whew! I just drove $\frac{6}{5}$ of a mile." It's clearer to say, "Maggie drove $1\frac{1}{5}$ miles." See what I mean? The second answer sounds better, and it gives you a better feeling for how far she actually drove—even though $\frac{6}{5}$ and $1\frac{1}{5}$ have exactly the same value.

Sometimes your homework will specifically ask you to put your answer in one form or another, but whenever it *isn't* specified, here's the bottom line: Use improper fractions while you are in the *middle* of working on a problem, since improper fractions are easier to work with. Only convert the fraction to a mixed number for your *final answer*.

QUICK NOTE! By the way, a special case of improper fractions are fractions that look like this: $\frac{4}{1}$, $\frac{8}{1}$, $\frac{33}{1}$. Whenever you have "1" as the bottom (denominator), the fraction is really a *whole number* in disguise. So $\frac{4}{1} = 4$, $\frac{8}{1} = 8$, and $\frac{33}{1} = 33$.

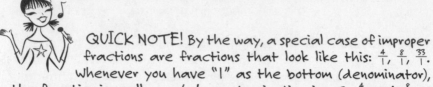

Doing the Math

What else do the following fractions equal? I'll do the first one for you.

1. $\frac{77}{1} = 77$

2. $\frac{6}{1} =$

3. $\frac{1}{1} =$

4. $\frac{141}{1} =$

(Answers on p. 287)

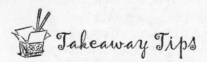

Takeaway Tips

- In a fraction, the *numerator* is on the top and the **denominator** is **d**own below the fraction line.

- *Improper fractions* and *mixed numbers* are two ways of writing fractions that have values of more than 1.

- *Improper fractions* have bigger numerators than denominators.

- *Mixed numbers* mix whole numbers with fractions.

- To convert improper fractions → mixed numbers, just tip the fraction over and divide.

- To convert mixed numbers → improper fractions, use the MAD Face!

Diamonds Are a Girl's Best Friend

As you probably know, diamonds are one of the most precious gems on the planet. It's that extra special "sparkle" they have that sets them apart. The reason they're so sparkly is because of their hardness and transparency, which allows light to travel through them very quickly.

Diamonds have this hardness and transparency because of the way they were formed: millions of years ago, deep under the earth's crust, struggling under extreme heat and pressure. All that struggling under extremely hard conditions is *what made them* so sparkly.

So why am I talking about diamonds? Okay, yes, because I really like diamonds. But also because, when you're having a hard time with math and you're really struggling, I want you to think of yourself as a diamond, forming deep beneath the Earth's crust, knowing that all that struggling will someday make you, yes, very sparkly.

Believe it or not, struggling with math actually makes you smarter. It might not always feel that way, but it's true. It's in the *struggling* that you'll come to understand the concepts better. The more you struggle and reason things out in your head, the sharper your math skills will be. It's all about being determined and not shying away from things when they get hard.

Someday you'll be a sparkly and fabulous woman yourself, with a great head on her shoulders and a killer job. And most valuable things take time and true, genuine effort—that's the way it happens in nature with diamonds, and that's how it happens in life . . . with you!

"I think a lot of girls dumb themselves down for boys. I don't see the point. I'm smart, and I also have a boyfriend. Besides, the guys you have to dumb yourself down for don't make good boyfriends anyway." Elyssa, 17

Chapter 5

How Many Iced Lattes Can These Actors Drink?

Multiplying and Dividing Fractions . . . and Reciprocals

$\mathcal{M}$ultiplying fractions is really easy—in fact, you probably already know how to do it, so what's here is just a quick review. (We'll get to the actors and their lattes in a minute.)

Multiplying Fractions

To multiply two fractions, simply multiply across the top and across the bottom. Voilà!

For example, $\frac{1}{3} \times \frac{2}{5}$

$$\frac{1}{3} \times \frac{2}{5} = \frac{1 \times 2}{3 \times 5} = \frac{2}{15}$$

So, $\frac{1}{3} \times \frac{2}{5} = \frac{2}{15}$ and that's your answer!

Check this out: Let's say you're in charge of hairdos for your school musical. Yeah, yeah. You tried for the lead part but forgot the words in the middle of your song because that guy you have a crush on was watching you. How embarrassing!

Anyway, now you're in charge of the hair, and you're determined to do a great job, because that's the kind of gal you are. So, there's a musical number with 5 girls and they all need to wear the same blue ribbon in their hair. Each ribbon should be $2\frac{1}{4}$ feet long. How much total ribbon should you get at the fabric store?

The girl in charge thinks you can't do math and are going to buy too much ribbon. We'll just show her, won't we?

We need to figure out $5 \times 2\frac{1}{4}$. *Hmm*. Well, we know how to multiply two fractions together, so let's express these two numbers as fractions!

First, let's convert 5 into an improper fraction: $5 = \frac{5}{1}$. Next, we'll convert $2\frac{1}{4}$ into an improper fraction, by using the MAD Face method we learned on p. 45.

$$+\overset{\curvearrowright}{\underset{\times}{(2\,\frac{1}{4}}}$$

M: <u>M</u>ultiply: $4 \times 2 = 8$

A: <u>A</u>dd: $8 + 1 = 9$

D: Put 9 over the <u>d</u>enominator: $\frac{9}{4}$

Now, instead of $5 \times 2\frac{1}{4}$, our problem looks like $\frac{5}{1} \times \frac{9}{4}$. (Remember, even though the numbers look different, they have the exact same value as they did before.) Let's see what we get when we multiply these fractions together:

$$\frac{5}{1} \times \frac{9}{4} = \frac{5 \times 9}{1 \times 4} = \frac{45}{4}$$

So, we know we need to get $\frac{45}{4}$ feet of ribbon.

But let's go ahead and convert it back into a mixed number, since it's a word problem. After all, it might be awkward to walk into a fabric store and say, "Hi, I'd like $\frac{45}{4}$ feet of blue ribbon, please!" Trust me, it wouldn't go over very well.

So, to turn an improper fraction into a mixed number, we just tip over the fraction and divide the top number by the bottom one, like we did on p. 42. And the remainder of 1 just becomes the top of the fractional part of the mixed number:

$$\frac{45}{4} = 4\overline{)45}^{\,11\ \text{R1}} \qquad\qquad \frac{45}{4} = 11\frac{1}{4}$$

Ah, much better. "Hi, I'd like $11\frac{1}{4}$ feet of blue ribbon, please!" Now that would work just fine.

As you can see, the most time-consuming part of multiplying fractions is converting mixed numbers into improper fractions first. The multiplying itself is easy, as long as you remember your times tables!

Step-By-Step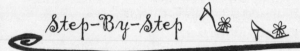

Multiplying fractions:

Step 1. Convert mixed numbers (if any) to improper fractions.

Step 2. Convert whole numbers (if any) to improper fractions.

Step 3. Multiply across the top and bottom.

Step 4. Reduce, if possible. Done!

Doing the Math

Multiply the following fractions. I'll do the first one for you.

1. $\frac{2}{3} \times \frac{4}{5} = \frac{2 \times 4}{3 \times 5} = \frac{8}{15}$

Answer: $\frac{8}{15}$

2. $\frac{1}{2} \times 3 =$

3. $1\frac{3}{7} \times \frac{1}{3} =$

(Answers on pp. 207–0)

Reciprocal Fractions

Sometimes it's helpful to turn fractions upside down. Yes, upside down. "Huh?" you may ask, "I'm not much a fan of fractions as it is, so why do you think I'd like them more upside down?!" Really, it's not so bad—and besides, they make dividing fractions much easier—but we'll get to that later. For now, it's time to flip!

When you flip a fraction upside down, the new fraction is called its **reciprocal**. For example:

$$\text{the reciprocal of } \tfrac{2}{5} \text{ is } \tfrac{5}{2}$$

$$\text{the reciprocal of } \tfrac{4}{7} \text{ is } \tfrac{7}{4}$$

$$\text{the reciprocal of } \tfrac{1}{3} \text{ is } \tfrac{3}{1}$$

When you see the word *reciprocal*, think "re-FLIP-ro-cal." Say it out loud right now. Really, say it: "re-FLIP-ro-cal." It's silly, but you know what? It'll help you remember what the heck a reciprocal is, so stop complaining!

What's It Called?

Reciprocal

The *reciprocal* of a fraction is found by flipping it upside down. If you want the reciprocal of a mixed number or a whole number, just convert it to an improper fraction, and then flip it!

QUICK NOTE! Because you can rewrite whole numbers as fractions (like: $5 = \tfrac{5}{1}$), the reciprocal of whole numbers is easy to find. What do you think the reciprocal of 5 is? You guessed it: $\tfrac{1}{5}$.

Step-By-Step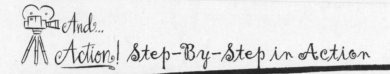

Finding a number's reciprocal:

Step 1. If it's a mixed number or a whole number, convert it to an improper fraction.

Step 2. Flip it upside down!

And... Action! Step-By-Step in Action

Let's find the reciprocal of $3\frac{1}{2}$.

Step 1. First, we'll convert it to an improper fraction using the MAD Face method (see p. 45):

M: $2 \times 3 = 6$
A: $6 + 1 = 7$
D: Put 7 over the denominator, $\frac{7}{2}$. So, $3\frac{1}{2} = \frac{7}{2}$.

Now we can find its reciprocal.

Step 2. Flip $\frac{7}{2}$ to get $\frac{2}{7}$.

So, the reciprocal of $3\frac{1}{2}$ is $\frac{2}{7}$. That's not so bad, right?

QUICK NOTE! Taking the reciprocal of a number is one of those mathematical actions where *if you do it twice, you'll end up right where you started.*

Say you're holding a magazine, and you flip it upside down, and then flip it upside down again. It will be right side up again, right?

The same logic applies to fractions. The reciprocal of $\frac{3}{8}$ is $\frac{8}{3}$, and the reciprocal of $\frac{8}{3}$ is $\frac{3}{8}$.

Another interesting thing to note about reciprocals is that if you multiply a number times its reciprocal, you'll always get an answer equal to 1. For instance, using our example from above, $\frac{3}{8} \times \frac{8}{3} = \frac{24}{24} = 1$. Try it with another example, you'll see that it works!

Doing the Math

Find the reciprocal of the improper fractions and mixed numbers below. I'll do the first one for you.

1. $1\frac{2}{5}$

<u>Working out the solution</u>: First let's convert it into an improper fraction.

M: $5 \times 1 = 5$

A: $5 + 2 = 7$

D: Put 7 over the denominator, $\frac{7}{5}$

Now we can flip it! $\frac{7}{5} \rightarrow \frac{5}{7}$

<u>Answer</u>: $\frac{5}{7}$

2. $\frac{8}{3}$

3. $2\frac{1}{2}$

4. $\frac{19}{296}$

5. 9

(Answers on p. 288)

Remember when I said that reciprocals would come in handy for division?

Dividing Fractions

Dividing fractions is almost as easy as multiplying them. When you divide one fraction by another fraction, you simply find the reciprocal of the second fraction (flip it!) and then *multiply* the two fractions together. How would we divide $\frac{3}{4} \div \frac{1}{5}$?

$$\frac{3}{4} \div \frac{1}{5} = \frac{3}{4} \times \frac{5}{1} = \frac{3 \times 5}{4 \times 1} = \boxed{\frac{15}{4}}$$

flip it!

So $\frac{3}{4} \div \frac{1}{5} = \frac{15}{4}$.

Step-By-Step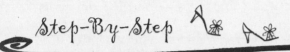

Dividing fractions:

Step 1: Make sure all whole numbers and mixed numbers are written as improper fractions.

Step 2. Flip the second fraction upside down (re-FLIP-rocal).

Step 3. Change the division sign to a multiplication sign.

Step 4. Multiply across the top and bottom, done!

But this seems like a math "trick"—why should that work? Why is multiplying by the reciprocal *the same as* dividing?

Let's look at the easy problem $10 \div 2 = 5$. What would happen if, instead of dividing by 2, we *multiplied* by the *reciprocal* of 2? The reciprocal of $2 = \frac{1}{2}$, so the problem would become:

flip it!

$$10 \div 2 = \frac{10}{1} \div \frac{2}{1} = \frac{10}{1} \times \frac{1}{2} = \frac{10 \times 1}{1 \times 2} = \frac{10}{2} = \boxed{5}$$

Yep, we got 5, so it works! And you can also see by this example that it's always the *second* number that has to get flipped. If you flipped the first number, it wouldn't work.

Let's do another example together. How about $4 \div \frac{1}{2}$?

flip it!

$$4 \div \frac{1}{2} = \frac{4}{1} \div \frac{1}{2} = \frac{4}{1} \times \frac{2}{1} = \frac{4 \times 2}{1 \times 1} = \frac{8}{1} = \boxed{8}$$

As you can see, our answer is 8.

But "8" is kind of an unexpected answer, isn't it? Seems strange that we would get such a big number when we're working with fractions. It makes sense that $4 \times \frac{1}{2} = 2$, because this means the same thing as $4 \div 2$, right?

But $4 \div \frac{1}{2} = \mathbf{8}$? Let's think about this for a second.

What does it "mean" to divide by a fraction? Well, first let's remember what it means to "divide" . . .

You're back at that musical, and now that everyone has their precious blue ribbons, you are sent on a latte run. Yep, they want *you* to go grab lattes for all the lead actors. You come back with 6 ice-blended lattes—that's all you could carry—and the lead actors demand 2 lattes each.

How many people could you serve? Well, if you have 6 lattes, and each person wants 2 of them, you'd have to divide 6 by 2: $6 \div 2 = 3$. So, you could serve 3 people their beloved coffee drinks.

Remember, however, that this division problem $(6 \div 2)$ is really asking the question, "How many times does 2 go into 6?" Or, saying the same thing in different words, "How many 2s are there in 6?"

When we do $6 \div 1$, we get 6. That's because that division problem asks the question, "How many times does 1 go into 6?" We know that 1 goes into 6 exactly 6 times. If the actors were a bit more reasonable and only demanded 1 latte each, you could serve a total of 6 people: $6 \div 1 = 6$. That's a pretty easy division problem, huh?

Keeping this in mind, consider this next question: "How many times does $\frac{1}{2}$ go into 6?" This time, you still have a total of 6 ice-blended lattes, but you've decided that these picky actors are only going to get *half* a latte each. After all, now you can serve more of the cast, right? You're just trying to do your job! So, that would be $6 \div \frac{1}{2}$, right?

How many times do *you* think that $\frac{1}{2}$ goes into 6?

Well $\frac{1}{2}$ is pretty small, smaller than 1, so we know that it probably goes into 6 a bunch of times. You could always count it out:

$$1 + 1 + 1 + 1 + 1 + 1 = 6$$
$$\underbrace{\tfrac{1}{2}+\tfrac{1}{2}} + \underbrace{\tfrac{1}{2}+\tfrac{1}{2}} + \underbrace{\tfrac{1}{2}+\tfrac{1}{2}} + \underbrace{\tfrac{1}{2}+\tfrac{1}{2}} + \underbrace{\tfrac{1}{2}+\tfrac{1}{2}} + \underbrace{\tfrac{1}{2}+\tfrac{1}{2}} = 6$$

As it turns out, $\frac{1}{2}$ goes into 6 exactly 12 times. In other words, "There are twelve $\frac{1}{2}$s in 6." And in math words: $6 \div \frac{1}{2} = 12$. So you could serve 12 actors lattes, if they were willing to just get half a latte each. Wishful thinking, right?

Now, the easiest and quickest way to get the answer when you're dividing fractions is to simply use the multiplication/reciprocal method of dividing we learned above—but I showed you the above example because I wanted to help the answers *make more sense*, since the concept of "dividing by a fraction" can be hard to wrap your head around.

Doing the Math

Divide the following fractions. I'll do the first one for you.

1. $2 \div \frac{1}{100} =$

<u>Working out the solution:</u>

$$2 \div \frac{1}{100} = \frac{2}{1} \div \frac{1}{100} = \overset{\curvearrowleft \text{flip it!}}{\frac{2}{1} \times \frac{100}{1}} = \frac{2 \times 100}{1 \times 1} = \frac{200}{1} = \boxed{200}$$

<u>Answer:</u> 200. This makes sense because you could fit a lot of $\frac{1}{100}$s into 2!

2. $\frac{4}{5} \div \frac{3}{2} =$

3. $3\frac{1}{5} \div 5 =$

4. $\frac{1}{2} \div 3 =$

(Answers on p. 288)

Takeaway Tips

- *Multiplying fractions* is easy: Just multiply across the top and bottom and you're done!

- To find the *reciprocal* of a fraction, simply flip the fraction upside down. reciprocal = re-FLIP-rocal!

- When *dividing fractions*, flip the *second* fraction upside down, then *multiply* the two fractions together to get your answer. If you ever get confused about which fraction to flip, just think about the 10 ÷ 2 problem (see p. 57).

It's funny how people's expectations of you can dictate how you feel about yourself. If your mom expects you to pick up your room, you will probably find yourself keeping your room neat. If you know that no one is watching, and no one expects you to be neat, I bet your room gets pretty messy.

It's normal—it's human nature—but we have to be careful that we don't fall victim to other people's low expectations!

I'll never forget what happened in my ninth grade science class. After our first test, my science teacher pulled me aside and expressed surprise at my high score, exclaiming how unexpected it was that I would be a good student in science. "You just seem so outgoing and you wear such brightly colored earrings. I just didn't think you would be very smart."

Can you believe it? Somehow she thought that just because I was socially well-adjusted and cared about how I looked, I wouldn't be intelligent. I was floored. All based on appearances! The teacher was judging me according to the stereotypes that are so deeply ingrained in our society.

I remember thinking, "I'll show *her* I don't have to look like a dork to be smart!" I mean, who did she think she was? She probably had no idea how backward her thinking was or how it might have affected me in a negative way. Luckily, it had the opposite effect on me because it made me so mad.

I did not live up to her low expectations. I got an A in that class and continued to wear big, fun earrings. I could have smarts *and* a flair for fashion. You don't have to choose one or the other.

In fact, when you think of the bright, fabulous woman you want to be someday, remember that you are in training for that *right now*.

Imagine yourself as a high-powered attorney or the owner of a fashion magazine—wearing your expensive suit and high heels, and carrying your briefcase full of important work for the day. Sound good? Every challenge you overcome *today* will bring you closer to that goal. Every homework problem you think you can't do—but then through determination, you solve—every time you exercise your brain and your beauty, inside and out, you're becoming the young woman you aspire to be. I'm here to tell you from personal experience that you can be a glamour girl *and* a smart young woman—who can *certainly* do math.

Chapter 6

When to Seriously Stop Raiding the Refrigerator

Equivalent Fractions and Reducing Fractions

*I*n the English language, there are many ideas that can be expressed in different ways, using different words. For example, if you really like your friend's blouse, you could say, "That's a cute *blouse*." Or you could say, "That's a cute *top*." Then you might say, "Should I wear *nylons* with this dress?" Or, "Should I wear *pantyhose* with this dress?" On another occasion, you might say, "That guy is a *jerk*," or "That guy is a *total scumbag loser*" . . . Okay, that one has a bunch of options.

See what I mean? There's almost always more than one way to say the same thing.

In the language of math, there are *lots* of ways to express the same number, especially when it comes to fractions. In chapter 4, you saw how this is true for improper fractions and mixed numbers—although they look different, both can have the same value. In this chapter, we'll see how this relates to equivalent fractions and to reducing fractions. But first, to demonstrate this point, let me tell you a story about me . . . and pie.

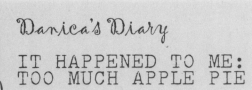

Danica's Diary

IT HAPPENED TO ME: TOO MUCH APPLE PIE

One Thanksgiving a few years ago, my mom, who is an excellent cook, baked one too many pies for our guests. She made a pumpkin pie, a pecan pie, and an apple pie—all sweetened with just honey, no sugar. Yum! But the apple pie didn't even get touched, everyone was so full from the others.

The next day, the untouched apple pie was just sitting in our fridge, and boy did it look good. After lunch, I cut a thin little slice, thinking to myself, "I'm not eating very much—look at how small this slice is!" A few minutes later, I went back and had another little slice—then another, and another.

Pretty soon I noticed that *half the pie* was gone! At about the same time, I began to feel really sick to my stomach. I never would have cut myself an entire *half of a pie*, plopped it on a plate, and eaten the whole thing! But as much as I hated to admit it, that's essentially what I had done.

Each time I cut a slice, it was around $\frac{1}{12}$ of the pie (see picture above). And how many of those slices did it take to make a half a pie? Just 6! This is an excellent demonstration of how $\frac{6}{12} = \frac{1}{2}$.

So, $\frac{6}{12}$ and $\frac{1}{2}$ are **equivalent fractions**. They may look different, they may use different numbers, but they represent the *same value*—and when you're talking about pie, both will make you just as sick to your stomach. Saying two fractions are "equivalent" to each other is just a fancy way of saying they are "equal in value." You probably wouldn't want to eat an entire half of a pizza, either—no matter how you cut it up.

There are other ways to cut $\frac{1}{2}$ of a pizza: like $\frac{4}{8}$ and even $\frac{11}{22}$, if you had a sharp enough knife. Cutting $\frac{1}{2}$ a pizza into $\frac{4}{8}$ would give you 4 little slices, where each is $\frac{1}{8}$ of a pizza. And cutting $\frac{1}{2}$ a pizza into $\frac{11}{22}$ would give you 11 *really* tiny slices, where each is $\frac{1}{22}$ of a pizza. These amounts are both equivalent to $\frac{1}{2}$ of a pizza. See below how $\frac{4}{8} = \frac{11}{22} = \frac{1}{2}$.

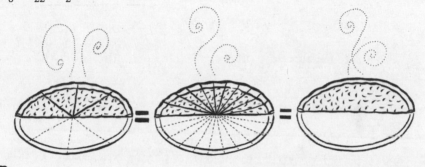

What's It Called?

Equivalent Fractions

Two fractions are considered equivalent when they represent the same value—the same amount of pie. For example, $\frac{1}{2}$ and $\frac{2}{4}$ are *equivalent fractions*. So are $\frac{1}{2}$ and $\frac{32}{64}$. (In other words, if you were to eat $\frac{1}{2}$ of a pie or $\frac{32}{64}$ of a pie, you'd be eating the same amount.)

Finding Equivalent Fractions

If I told you I was going to eat $\frac{1}{4}$ of a pizza, but I wanted to cut the pizza into thin slices, what are some options for the sizes of slices (expressed as fractions) that I could eat?

Well, I might divide the whole pizza into 8 slices. In that case, $\frac{1}{4}$ would equal $\frac{1}{8} + \frac{1}{8}$ of the pizza, which is $\frac{2}{8}$. Notice that if you multiply the top and bottom of $\frac{1}{4}$ by 2, you'll get this equivalent fraction: $\frac{1 \times 2}{4 \times 2} = \frac{2}{8}$

In fact, if you take $\frac{1}{4}$ and multiply *any* number by both the bottom *and* the top of the fraction, you'll end up with a fraction that is *equivalent* to $\frac{1}{4}$. Remember, equivalent fractions are just two ways of expressing the *same amount* (of pizza, or pie, or whatever!). For instance:

$\frac{1}{4}$ is the same as $\frac{1 \times 2}{4 \times 2} = \frac{2}{8}$ (*we multiplied top and bottom of $\frac{1}{4}$ by* **2**)

same as $\frac{1 \times 3}{4 \times 3} = \frac{3}{12}$ (*we multiplied top and bottom of $\frac{1}{4}$ by* **3**)

same as $\frac{1 \times 4}{4 \times 4} = \frac{4}{16}$ (*we multiplied top and bottom of $\frac{1}{4}$ by* **4**)

same as $\frac{1 \times 5}{4 \times 5} = \frac{5}{20}$ (*we multiplied top and bottom of $\frac{1}{4}$ by* **5**).

The above fractions are all *equivalent* to each other—they all represent the same *value*: $\frac{1}{4}$.

QUICK NOTE! Remember, when two fractions are *equivalent*, it means that the two fractions may look different, but they represent the *same value*—the same amount of pizza or pie.

Copycat Fractions

A long time ago, you probably learned that <u>multiplying a number by 1 doesn't change the number</u>. For instance, $4 \times 1 = 4$, $\frac{1}{7} \times 1 = \frac{1}{7}$, $492 \times 1 = 492$, etc. It doesn't matter *what* number you multiply by 1, the simple fact is this: Multiplying any number by the number 1 has *no effect* on its value.

You may also have heard that when a fraction's numerator and denominator are the same, then the fraction equals 1. So all fractions that look like $\frac{2}{2}$, $\frac{8}{8}$, $\frac{39}{39}$, or $\frac{1492}{1492}$ each equal 1.

I like to call these fractions **copycat fractions**. It's just a name I came up with, because, I mean, the top and the bottom are like copycats—they're the exact same thing. Your math teacher probably just says to "multiply top and bottom by the same number" to get an equivalent fraction, but in reality, this is the same thing as multiplying by a copycat fraction. (And I happen to think that copycat fractions are easier to remember!)

You can make your very own copycat fraction by putting any whole number (except zero) on the top and bottom of a fraction. And remember: **The value of every "copycat fraction" is 1!**

Naturally, then, since all these copycat fractions are equal to 1, we can multiply them by any number without changing the value of the number. So when you multiply a copycat times a fraction, you end up with an equivalent fraction—it'll look different, but it has the *same value!*

$$\text{fraction} \times \frac{\textbf{copycat}}{\textbf{copycat}} = \text{equivalent fraction}$$

Take the fraction $\frac{1}{4}$. Multiply it by any copycat, let's say $\frac{2}{2}$, and you'll get an equivalent fraction: $\frac{1}{4} \times \frac{2}{2} = \frac{2}{8}$. In fact, we could rewrite the table on page 64 and see that it actually showed a bunch of **copycats** at work, giving us our equivalent fractions.

$$\frac{1}{4} = \frac{1}{4} \times \frac{\textbf{2}}{\textbf{2}} = \frac{1 \times 2}{4 \times 2} = \frac{2}{8} \text{ (multiplied by copycat } \tfrac{\textbf{2}}{\textbf{2}}\text{)}$$

$$\frac{1}{4} = \frac{1}{4} \times \frac{\textbf{3}}{\textbf{3}} = \frac{1 \times 3}{4 \times 3} = \frac{3}{12} \text{ (multiplied by copycat } \tfrac{\textbf{3}}{\textbf{3}}\text{)}$$

$$\frac{1}{4} = \frac{1}{4} \times \frac{\textbf{4}}{\textbf{4}} = \frac{1 \times 4}{4 \times 4} = \frac{4}{16} \text{ (multiplied by copycat } \tfrac{\textbf{4}}{\textbf{4}}\text{)}$$

$$\frac{1}{4} = \frac{1}{4} \times \frac{\textbf{5}}{\textbf{5}} = \frac{1 \times 5}{4 \times 5} = \frac{5}{20} \text{ (multiplied by copycat } \tfrac{\textbf{5}}{\textbf{5}}\text{)}$$

Watch Out! The expression $\frac{0}{0}$ is *not* a copycat fraction, because $\frac{0}{0}$ is undefined in math. You'll *never* see $\frac{0}{0}$; it just doesn't have a value. In fact, you can never have 0 on the bottom of a fraction—ever. Seriously. I'm not even kidding.

Doing the Math

Rewrite the following fractions by multiplying them times the copy-cat fractions $\frac{2}{2}$, $\frac{3}{3}$, and $\frac{10}{10}$. I'll do the first one for you.

1. $\frac{2}{7} =$

<u>Working out the solution</u>:

$$= \frac{2}{7} \times \frac{2}{2} = \frac{2 \times 2}{7 \times 2} = \frac{4}{14}$$

$$= \frac{2}{7} \times \frac{3}{3} = \frac{2 \times 3}{7 \times 3} = \frac{6}{21}$$

$$= \frac{2}{7} \times \frac{10}{10} = \frac{2 \times 10}{7 \times 10} = \frac{20}{70}$$

<u>Answer</u>: Some equivalent fractions to $\frac{2}{7}$ are $\frac{4}{14}$, $\frac{6}{21}$, and $\frac{20}{70}$.

2. $\frac{1}{2} =$

3. $\frac{4}{3} =$

4. $5 =$ (Hint: Rewrite 5 as $\frac{5}{1}$ and then you can treat it like a fraction!)

(Answers on p. 288)

"*If you actually pay attention, math is pretty easy.*" Alexis, 13

"*I used to think that math was just a waste of time, but I now see that math is really important—for science, music, school, and just everyday life.*" Brianna, 12

Reducing Fractions

After you've solved a math problem (like a word problem) that involves fractions, the instructions will usually say to leave your answer in "reduced form" or "simplest form." Or they'll say, "Reduce your answer." That just means that they want the fraction to be expressed using *the smallest numbers possible*.

In other words, when you are asked to **reduce** a fraction, you are being asked to find the *smallest numbers* you can use to express the *same amount*. Smaller numbers are just easier to deal with, y'know? I'd rather someone show me a nice and simple-looking, *reduced* fraction like $\frac{1}{2}$ than a number like $\frac{32}{64}$.

But remember: $\frac{1}{2}$ and $\frac{32}{64}$ have the *same value*. They're *equivalent fractions*. They represent the *same* amount of pizza.

QUICK NOTE! When you *reduce* a fraction, the *reduced fraction* is an *equivalent fraction* to the one you started off with. It makes sense that the two fractions would be equivalent, because you haven't changed the *value* of the fraction by reducing it. As you know, equivalent fractions are fractions that use different numbers but have the same value!

"How exactly do you convert fractions that use 'big numbers' to fractions that use 'small numbers'"? I'm glad you asked.

Reducing Fractions = Taking out the Hidden Copycats

Remember how we multiplied fractions by copycats to get *equivalent fractions*? Like:

$$\frac{2}{3} = \frac{2 \times \mathbf{4}}{3 \times \mathbf{4}} = \frac{8}{12}$$

Reducing fractions is just going in reverse: Instead of *multiplying* a fraction by a copycat, what we want to do is to seek out the *hiding copycat fractions*, then *divide* them away, as if they were never even there.

For example, let's reduce the fraction $\frac{8}{12}$.

The first step is to see if the top and bottom have any of the same *factors*, because if they do, there is a copycat fraction hiding inside! In order to reduce fractions, we have to be detectives: we're going to hunt down the hiding copycat fractions and erase all evidence of them being there.

Since 4 is a factor of 8 and 12, we know we can express 8 and 12 like this:

$$8 = 2 \times 4$$
$$12 = 3 \times 4$$

Then we can substitute "2×4" for 8 and "3×4" for 12 and rewrite $\frac{8}{12}$ as $\frac{2 \times 4}{3 \times 4}$.

$$\frac{8}{12} = \frac{2 \times 4}{3 \times 4}$$

Aha! Well, lookie there. It seems as if once upon a time the innocent little fraction $\frac{2}{3}$ was multiplied by the sneaky copycat fraction $\frac{4}{4}$, and we wound up with $\frac{8}{12}$, which has the same value as $\frac{2}{3}$ but is expressed in a form that has *way* too many big numbers for our taste . . .

So, once upon a time, this happened: $\frac{2}{3} = \frac{2 \times 4}{3 \times 4} = \frac{8}{12}$. And now, going *backward*, we can do this: $\frac{8}{12} = \frac{8 \div 4}{12 \div 4} = \frac{2}{3}$.

You may not realize it, but we just *reduced* $\frac{8}{12}$ to its simplest form, $\frac{2}{3}$.

What's It Called?

Reduced Form

A fraction is in *reduced form* when the numerator (top) and denominator (bottom) share no common factors. It's also sometimes called "simplest form," "reduced terms," or "simplest terms."

Dividing vs. "Canceling"

The most common way to reduce fractions is to *divide* the top and bottom by a common factor (i.e., to hunt down copycats like a sleuth detective, then divide them out!), like we did in the last example:

$$\frac{8}{12} = \frac{8 \div 4}{12 \div 4} = \frac{2}{3}$$

Sometimes, though, you'll be told to **cancel** the common factors on the top and bottom of the fraction.

For instance, in the case of $\frac{8}{12}$, you might rewrite the top and bottom as factors and *cancel* the common ones, or even just do it in your head:

$$\frac{8}{12} = \frac{2 \times \cancel{4}^{\,1}}{3 \times \cancel{4}_{\,1}} = \text{ or even just: } \frac{\cancel{8}^{\,2}}{\cancel{12}_{\,3}} = \frac{2}{3}$$

I love canceling. Who doesn't? It feels fantastic. It makes the numbers smaller, and everything just looks neater—it's almost like you're canceling out the math homework you have to do!

But it's helpful to understand *why* you can cancel. Otherwise, we could all get so "cancel happy" that we just start putting lines through every number we see. Hah! I hereby *cancel* you . . . ! See how we could get into all sorts of trouble like that?

Watch Out! We can multiply or divide the top and bottom of a fraction by the same number without changing its value, but we *can't add or subtract* a number from the top and bottom, or we'd change the value of the fraction. For example: $\frac{3}{6} \neq \frac{3-3}{6-3} = \frac{0}{3}$. (This would change the value of the fraction—don't do this!)

In fact, the only reason you're allowed to *divide* the top and bottom by the same number without changing the value of the fraction is because you're essentially walking backward through the process of multiplying by a copycat fraction.

We know that multiplying by a copycat fraction (whose value is always 1) does not change the value of the fraction (see p. 65). So it makes sense that going *backward* through that process should have no effect on its value, either.

If you have a pair of sunglasses and you fold out the sides, it doesn't change the "value" of the sunglasses. So, if you went *backward* through that process, and you folded the sides back in, it obviously wouldn't change their value, either.

Strategies for Reducing Fractions

Below are strategies to figure out *what* number to divide the top and bottom of a fraction by—because the hardest part of reducing fractions is figuring out what copycat fraction is hiding inside. The actual dividing (or canceling) is easy!

Strategy #1. The GCF Method. You can find the GCF (Greatest Common Factor) of the numerator (top) and denominator (bottom), and then divide both numbers by the GCF. See chapter 2 to review methods for finding the GCF. My favorite is the Birthday Cake method.

Strategy #2. The Divide and Conquer Method. Or, you can just hunt and find smaller common factors of the numerator (top) and denominator (bottom), then keep dividing the numerator and denominator by those factors until there are no more common factors left. The divisibility chart on page 9 is really helpful for finding common factors in this method (I recommend using it until you have memorized it), which I like to call the Divide and Conquer method, because it's like you're hacking away at the factors, slaying them, one by one, with your sword . . . or whatever.

Whichever method you prefer, both boil down to the same thing: figuring out which *common factors* are shared by the top and bottom of the fraction, and then dividing top and bottom by those factors to get rid of the hidden copycats.

 *Watch Out!* The word "reduce" makes it sound like you are reducing the *value* of the fraction, but that's *not* true: Reducing a fraction does *not* change the value of the fraction; it just simplifies how the fraction looks, by taking out as many copycat fractions (common factors) as possible.

Step-By-Step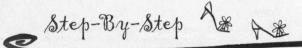

Reducing fractions using the GCF method:

Step 1. Find the GCF of the numerator (top) and denominator (bottom). (See chapter 2 for a review of finding the GCF.)

Step 2. Divide the numerator and denominator by the GCF. Done!

Reducing fractions using the Divide and Conquer method:

Step 1. Eyeball the numerator (top) and denominator (bottom) and see if they share any common factors.

Step 2. Divide by that common factor.

Step 3. Repeat until the top and bottom have no common factors left. Done!

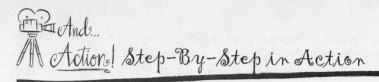

And...
Action! Step-By-Step in Action

Let's reduce $\frac{12}{30}$ by using the GCF method.

Step 1. To find the GCF, let's use the Birthday Cake method (see page 20).

$$\begin{array}{r|cc} 2 & 12 & 30 \\ \hline & 6 & 15 \end{array} \rightarrow \begin{array}{r|cc} 2 & 12 & 30 \\ 3 & 6 & 15 \\ \hline & 2 & 5 \end{array}$$

The two numbers on the left are 2 and 3, so the GCF is $2 \times 3 = 6$.

Step 2. Next, we divide the top and bottom by 6:

$$\frac{12}{30} = \frac{12 \div 6}{30 \div 6} = \frac{2}{5}$$

Since 2 and 5 share no common factors, we've successfully put the fraction in simplest, reduced form. I encourage you to use equal signs "=" when you're reducing, to remind yourself that all the fractions are *equal to each other*.

Take Two! Another Example!

Let's reduce $\frac{30}{54}$ using the Divide and Conquer method.

Step 1. Do the top and bottom have any factors in common? They're both even, so it looks like they have **2** in common.

Step 2. Okay, so now we'll divide both the top and bottom by **2**.

$$\frac{30}{54} = \frac{30 \div 2}{54 \div 2} = \frac{15}{27}$$

Step 3. Repeat until the top and bottom have no common factors left. Let's see, 3 goes into both 15 and 27, so let's try dividing the top and bottom by **3**.

$$\frac{15}{27} = \frac{15 \div 3}{27 \div 3} = \frac{5}{9}$$

Aha! Since 5 and 9 don't have any common factors, we're done!
Answer: $\frac{30}{54}$ in reduced form is $\frac{5}{9}$.

As you can see, the Divide and Conquer method takes a little longer than the GCF method, but, especially if you're starting with really big numbers, it can help you get started.

Take, for instance, $\frac{128}{192}$. This one might go a little something like this.

"Hmm, well, I feel a little lazy, so, since both numbers are even, I'm just going to start by dividing the top and bottom by 2."

$$\frac{128}{192} = \frac{128 \div 2}{192 \div 2} = \frac{64}{96}$$

"Hey, look! They're both even again. I guess I'll divide top and bottom by 2 again."

$$\frac{64}{96} = \frac{64 \div 2}{96 \div 2} = \frac{32}{48}$$

"Okay, okay. Yes they are both still even, but now from my times tables, I recognize that 32 and 48 are both divisible by 8."

$$\frac{32}{48} = \frac{32 \div 8}{48 \div 8} = \frac{4}{6}$$

"Well, I guess they're still both even, which means there's *still a* copycat fraction hiding in there, so let's take that one out now."

$$\frac{4}{6} = \frac{4 \div 2}{6 \div 2} = \frac{2}{3}$$

"Finally! Okay, 2 and 3 don't have any common factors. So we're done. The fraction $\frac{128}{192}$, in reduced form, is $\frac{2}{3}$."

All that, to find out that hiding underneath all those copycat fractions was sweet innocent little $\frac{2}{3}$? Yep, if you ate $\frac{128}{192}$ of a pie, you'd be eating $\frac{2}{3}$ of it. They're expressing the same value. You also could have found the GCF of 128 and 192 by using the GCF method (the GCF is 64, by the way), but that would have taken awhile, too.

Bottom line: My advice for big numbers? Don't use the GCF method for numbers bigger than say, 50. The list of factors simply gets too long. I'd hack away at the bigger numbers until they are both below 50, and then if you want to use the GCF method, go for it.

And, of course, there's always the golden math rule: since every problem is different, use whatever method seems easiest and fastest!

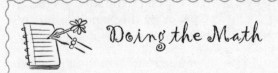

Doing the Math

Put these fractions in reduced form, using either the GCF method, the Divide and Conquer method, or a combination of both. I'll do the first one for you.

1. $\frac{48}{63}$

Working out the solution: Because these numbers are kind of big, I'll use the divisibility tricks from the chart on page 9 and notice that top and bottom are both divisible by **3**. So, $\frac{48}{63} = \frac{48 \div 3}{63 \div 3} = \frac{16}{21}$. Now what? Well, I know that the only factors of 21 are 3 and 7 (which are prime), and neither of those are factors of 16, so we're actually done!

Answer: $\frac{16}{21}$

2. $\frac{12}{18}$

3. $\frac{146}{168}$

4. $\frac{132}{165}$

(Answers on p. 288)

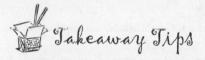 Takeaway Tips

- If two fractions are *equivalent*, it means they have the same value; they represent two ways of writing the exact same amount.

- Just like in the English language, in math, there are *lots of ways of writing the same value*, especially when it comes to fractions.

- To find an *equivalent fraction*, simply multiply a fraction by a *copycat fraction*.

- *Reducing fractions doesn't change their value.* After all, you're just walking backward through the process of multiplying by a copycat fraction (whose value is always 1).

- Multiplying or dividing by a *copycat fraction* is the same as multiplying or dividing the top and bottom by a *common factor*. It's just an easier way to think about common factors.

- *To reduce a fraction*, you divide the numerator (top) and denominator (bottom) of the fraction by their common factors—either all at once (using the GCF method) or by hacking away at them until there are no more common factors left (the Divide and Conquer method).

Chapter 7

Is Your Sister Trying to Cheat You Out of Your Fair Share?

Comparing Fractions

$\mathcal{S}$ay you and your sister both ordered your own pizzas, but neither of you could decide which toppings you wanted. (Ham and pineapple sounded pretty good, but veggie is your favorite, after all.) So you ordered veggie, and your sister ordered ham and pineapple, and you agreed to share.

But as soon as the pizzas arrived, your sister remembered that she'd promised her best friend she'd help her pick out an outfit for an upcoming party! Your sister now wants to split up the pizzas and take hers to her friend's house. When you open the pizza boxes, you find that even though they are the same size, the pizzas were *cut up* differently. Yours is cut into 8ths, and your sister's is cut into 6ths.

Your sister offers you 2 of her ham and pineapple pieces ($\frac{2}{6}$ of a pizza), and you offer her 3 of your veggie pieces ($\frac{3}{8}$ of a pizza). She then complains, saying you should give her 4 of your pieces, because your pieces are so much smaller. Is she right—or could she be trying to cheat you out of your fair share? Which is the bigger amount of pizza, anyway?

Sometimes it's easy to compare fractions. For instance, you might recognize right away that $\frac{1}{2}$ and $\frac{2}{4}$ are *equivalent*—that they are two ways of expressing the same value. And when two fractions have the same denominator—like $\frac{1}{3}$ and $\frac{2}{3}$, for example—you can simply look at the numerator to figure out which is bigger. Also, it's pretty clear that $\frac{1}{3}$ is bigger than $\frac{1}{8}$, right? After all, if you divided a pizza into 3 equal pieces and took one, that piece should be *bigger* than if you cut the pizza into 8 equal pieces and took one. But what about $\frac{1}{7}$ and $\frac{2}{11}$? Or $\frac{4}{6}$ and $\frac{34}{51}$? Hmm.*

What's It Called?

Symbols Review

$<$ means "is less than." For example, $\frac{1}{3} < \frac{2}{3}$

$>$ means "is greater than." For example, $\frac{1}{7} > \frac{1}{8}$

Just think of the $<$ and $>$ as hungry little alligator mouths—they always want to eat the bigger number! To this day, that's still how I tell them apart. You think I'm kidding, but I'm not.

Oh, and a circle like this $\bigcirc$ just represents the space waiting to be filled in with either $<$ or $>$, once the problem is done.

Comparing Fractions:
Using Copycats to Make the Denominators the Same

If you want to compare $\frac{1}{8}$ and $\frac{3}{8}$, it's easy to tell that $\frac{1}{8} < \frac{3}{8}$, since when both fractions have the same denominator (bottom), the numerators (tops) will tell you which fraction is bigger.

....

* Believe it or not $\frac{4}{6}$ and $\frac{34}{51}$ are equivalent fractions; they represent exactly the same amount of pizza!

But let's say you want to compare $\frac{1}{2}$ and $\frac{3}{8}$.

In this case, neither the numerators nor the denominators are the same, so it's not so easy to tell. What to do? *Multiply one of the fractions times a copycat (which we know won't change the <u>value</u> of the fraction), so that they will have the same denominator.*

In this case, let's multiply $\frac{1}{2}$ times the copycat fraction $\frac{4}{4}$, so that we'll end up with a denominator of 8:

$$\frac{1}{2} = \frac{1 \times 4}{2 \times 4} = \frac{4}{8}$$

Now that's much easier!

$$\frac{4}{8} > \frac{3}{8}$$
$$\downarrow \qquad \downarrow$$
$$\frac{1}{2} > \frac{3}{8}$$

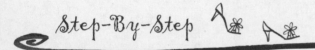

Step-By-Step

Comparing fractions with copycats:

Step 1. Make the denominators of both fractions the same, by multiplying one of the fractions by a copycat fraction.

Step 2. Compare the new numerators. This will tell you which fraction is bigger. Done!

And... Action! Step-By-Step in Action

Let's compare $\frac{5}{7}$ and $\frac{13}{21}$.

Step 1. We want the denominators to be the same, and we know that $7 \times 3 = 21$, so let's multiply $\frac{5}{7}$ by 3 on the top and bottom. To do this, we'll use the copycat fraction $\frac{3}{3}$.

$$\frac{5}{7} = \frac{5 \times 3}{7 \times 3} = \frac{15}{21}$$

Step 2. Ah, much easier now to see that $\frac{15}{21} > \frac{13}{21}$, so . . .

Answer: $\frac{5}{7} > \frac{13}{21}$

 Take Two! Another Example!

What if we wanted to compare two fractions like these: $\frac{5}{8} \bigcirc \frac{2}{3}$? There's no copycat fraction that takes us from one denominator to the other one—so we'll use *two* copycats, one on each fraction. Why not? It's a free country.

Step 1. In order to make these fractions' denominators the same, we look at the denominator from the *other* fraction and make a copycat fraction from it. Like, for $\frac{5}{8}$, we look at the other fraction's denominator, which is 3, and we multiply $\frac{5}{8}$ times $\frac{3}{3}$.

$$\frac{5}{8} = \frac{5 \times 3}{8 \times 3} = \frac{15}{24}$$

And for $\frac{2}{3}$, we look at the first fraction's denominator, which is 8, and we multiply $\frac{2}{3}$ times $\frac{8}{8}$.

$$\frac{2}{3} = \frac{2 \times 8}{3 \times 8} = \frac{16}{24}$$

So your page might look something like this:

$$\frac{3}{3} \times \frac{5}{8} \bigcirc \frac{2}{3} \times \frac{8}{8}$$
$$\rightarrow \frac{15}{24} \bigcirc \frac{16}{24}$$

Step 2. It's clear to see that $\frac{15}{24} < \frac{16}{24}$, so:

$$\downarrow \qquad \downarrow$$

Answer: $\dfrac{5}{8} < \dfrac{2}{3}$

 Doing the Math

Determine which fraction is bigger, or if they are equivalent, by first making their denominators the same (using copycats). I'll do the first one for you.

1. $\frac{3}{4} \bigcirc \frac{5}{7}$

<u>Working out the solution</u>: Since there's no copycat that can take us from 4 to 7, we'll need to use two copycat fractions, and change both of the fractions' denominators. We can get both denominators the same if we multiply $\frac{3}{4}$ by $\frac{7}{7}$ and if we multiply $\frac{5}{7}$ by $\frac{4}{4}$. That way, we'll make both of their denominators equal to $4 \times 7 = 28$:

$$\frac{7}{7} \times \frac{3}{4} \bigcirc \frac{5}{7} \times \frac{4}{4}$$

$$= \frac{21}{28} \bigcirc \frac{20}{28} \rightarrow \frac{21}{28} > \frac{20}{28}$$

<u>Answer</u>: $\frac{3}{4} > \frac{5}{7}$

2. $\frac{3}{4} \bigcirc \frac{4}{5}$

3. $2\frac{1}{3} \bigcirc \frac{21}{9}$

4. $\frac{5}{11} \bigcirc \frac{1}{2}$

(Answers on p. 288)

Shortcut Alert!
CROSS MULTIPLICATION
It's important to know how to compare two fractions by multiplying them by copycats to get their denominators the same, because you get more practice creating equivalent fractions, and mastering equivalent fractions will help you in tons of different types of math problems later on. But now it's time for me to show you a handy-dandy shortcut for situations when all you have to do is *compare* two fractions: **cross multiplication**.

Here's how it works: Write the two fractions out next to each other, then multiply "up" across the diagonal lines, and write the product up in the corners: these are called **cross products**. Then compare the two cross products, and whichever fraction has a bigger number over it is the bigger fraction!

Here's the example we did before:

$$\tfrac{1}{2} \bigcirc \tfrac{3}{8}$$

This time we'll cross multiply and compare the products we get:

$$\overset{8}{} \quad \overset{6}{} \\ \tfrac{1}{2} \otimes \tfrac{3}{8}$$

Since the 8 is bigger than the 6, we know that $\tfrac{1}{2} > \tfrac{3}{8}$. Pretty nifty, huh?

What's It Called?

Cross Multiplication

Cross multiplication is the act of multiplying up "across" the diagonal lines of two fractions. The products you get are called *cross products*. The bigger product tells us which is the bigger fraction. If the cross products are equal, then the two fractions are *equivalent*. For example, if we compare these fractions by finding their cross products . . .

$$\overset{30}{} \quad \overset{30}{} \\ \tfrac{2}{5} \otimes \tfrac{6}{15}$$

. . . we know that these fractions are equivalent, since their cross products both equal 30. So we can write: $\tfrac{2}{5} = \tfrac{6}{15}$.

Comparing two fractions using cross multiplication:

Step 1. Write the fractions out side-by-side. (Any mixed numbers should be expressed as improper fractions.)

Step 2. Cross multiply "up" the diagonal lines, and write the cross products over the fractions.

Step 3. Whichever fraction has the bigger product over it, is the bigger fraction. (And if the two products are the same, the fractions are equivalent.)

And...

Action! Step-By-Step in Action

Remember that wacky example on page 75, where I told you that $\frac{4}{6}$ and $\frac{34}{51}$ were equivalent? Well, let's just see about that. . . .

Steps 1 and **2.**

$$204 \qquad 204$$
$$\frac{4}{6} \bigotimes \frac{34}{51}$$

Voilà!

Step 3. Since the two products are the same, 204 = 204, we know that the fractions are equivalent!

Answer: $\frac{4}{6} = \frac{34}{51}$

By the way, 51 is one of those numbers that seems prime, but isn't. In fact, $51 = 3 \times 17$, and since $34 = 2 \times 17$, $\frac{34}{51}$ actually reduces to $\frac{2}{3}$!

QUICK NOTE! If you are comparing two fractions with cross multiplication and the numbers are kinda big (and you don't feel like doing all that long multiplication) you can always reduce one or both fractions *before* cross multiplying, without messing anything up. After all, reducing fractions doesn't change the fractions' values—and it's their *values* that you're trying to compare, right?

 Take Two! Another Example!

Let's do this comparison: $3\frac{2}{3}$ ◯ $\frac{34}{10}$. Which one is bigger?

Well, first we'll convert $3\frac{2}{3}$ into an improper fraction, because we know that improper fractions are much easier to work with when you want to "do" something with a mixed number. So let's use the MAD Face method (see p. 45) to get $3\frac{2}{3} = \frac{11}{3}$.

Step 1. Our new problem looks like $\frac{11}{3}$ ◯ $\frac{34}{10}$. We could cross multiply *now*, or first reduce the fraction on the right, so we don't get stuck with a bunch of long, messy multiplication. After all, both 34 and 10 are even, so we can divide top and bottom by **2**.

$$\frac{34}{10} = \frac{34 \div \mathbf{2}}{10 : \mathbf{2}} = \frac{17}{5}$$

Step 2. Now the problem looks like this: $\frac{11}{3}$ ◯ $\frac{17}{5}$. So we cross multiply "up" and get $5 \times 11 = 55$ and $3 \times 17 = 51$. And the products go above their fractions, at the end of the arrows, like this:

$$\overset{\displaystyle 55 \qquad 51}{\frac{11}{3} \bigotimes \frac{17}{5}}$$

Step 3. Since $55 > 51$, we know that: $\frac{11}{3} > \frac{17}{5}$, and so, writing them in their original form: $3\frac{2}{3} > \frac{34}{10}$.

 Doing the Math

Use cross multiplication to compare the following numbers. I'll do the first one for you.

1. $\frac{8}{9}$ ◯ $\frac{11}{12}$

<u>Working out the solution</u>: Since we can't reduce either fraction (they are both in their simplest form), let's just go ahead and cross multiply.

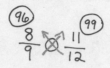

Since 96 < 99, we have our solution.

<u>Answer</u>: $\frac{8}{9} < \frac{11}{12}$

2. $\frac{20}{14}$ ◯ $\frac{90}{60}$ (Hint: reduce first!)

3. $\frac{1}{21}$ ◯ $\frac{1}{22}$

4. $\frac{100}{3}$ ◯ $\frac{3}{100}$

5. $\frac{17}{51}$ ◯ $\frac{1}{3}$

6. $\frac{2}{6}$ ◯ $\frac{3}{8}$

(Answers on p. 288)

Hmm, $\frac{2}{6}$ and $\frac{3}{8}$. Those fractions look familiar—they're the amounts of pizza you and your sister each gave the other at the beginning of the chapter.

But hold on a second—*was* your sister trying to cheat you out of your share of the pizza? Maybe she just doesn't understand fractions. (I can think of someone who can help her with that!)

Takeaway Tips

- *Comparing fractions is easiest when the denominators are the same.* If they're not, you can use copycat fractions to make them the same.

- *Cross multiplication* is when you multiply "up" the diagonal lines of two fractions and put the cross products over the fractions. The bigger cross product tells you which fraction is bigger!

- *Reducing fractions doesn't change their value,* and it usually makes the fractions easier and faster to compare. Sounds good to me.

"*Smart girls definitely have an edge. They demand a certain amount of respect.*" Rich, 18

"*I think smart girls are the ones with the most options in life—they can choose to be businesswomen, superstars, or homemakers, because they know how to do most anything.*" Iris, 15

"*I hate it when girls act dumb because they are afraid of showing what they can really do.*" April, 15

TESTIMONIAL:

Jessica Tan (New York, NY)

<u>Before</u>: High School study buddy
<u>Today</u>: Wall Street Whiz

When I was in middle school, math was my favorite subject. My dad taught me that math gives you an advantage in so many areas, from comparing prices while shopping to playing card games. Whether it was using basic math to help budget my allowance or using fractions to figure out how to evenly split a birthday cake, I realized that math was everywhere, and that being good at it made so many other things easier.

But not *everything* was easy. One day in my geometry class, a girl called me a "nerd" because I got a better score than she did on an exam. I was so upset! I didn't understand why anyone would penalize me for being smart. Years later, the same girl was in my high school calculus class.

There, she confessed to me that she was always jealous of anyone who was good at math, because she had always had a hard time with the subject. She apologized for being immature back in middle school (and calling me a nerd) and even asked if we could study together for an upcoming exam.

Looking back, I am so fortunate I had the courage not to shy away from something important just because somebody once called it "nerdy." And I am so glad I never stopped working hard at math because it led me to my current job, which I love.

> Being smart is something no one can ever take away from you.

Today, I'm in sales at an investment bank. I work with the stock market, which is an exchange where you can buy and sell shares of stock in different companies, and it's really fun!

If you buy stock in a corporation, you become a partial owner of that company. For example, say there is a company called Rockin' Records. If the entire Rockin' Records company is worth $7,000,000,000 and I own 100 shares at $80 per share, then I own 0.00011% of the company: $\frac{\$80 \times 100}{\$7,000,000,000} = 0.00011\%$.

Then, if I think that the price of Rockin' Records stock will increase by 20% in the next year, I would estimate that each share will eventually be worth $96, by calculating $80 × 1.20 = $96. Armed with this knowledge, I can then decide whether or not to invest in Rockin' Records!

Being smart is something no one can ever take away from you. I was really lucky that my parents always encouraged me to study hard, because not all math topics are easy—and I'm so glad I understand them today!

Chapter 8

How Much Do You and Your Best Friend Have in Common?

Common Denominators . . . and Adding and Subtracting Fractions

When the denominators of fractions are the same, we can simply add or subtract across the tops of these fractions. Let's talk a little more about this.

Common Denominators

My best friend Kimmie and I have a lot in common. We both care about the environment; we both love music, movies, arts and crafts; and we both have brown hair. (Well, that last one probably isn't very important!)

The most important stuff that we share in common is the stuff that makes each of us who we are, *deep down*: our morals, the way we view the world and our role in it, and so on. This makes sense, right? All those "deep down things" we share probably explain why we get along so well.

Here are two fractions that have a lot in common, "deep down"— that is, down below the fraction line.

$$\frac{1}{8} \text{ and } \frac{2}{8}$$

Do you see what they have in common? That's right—they have the same denominator. And when two fractions have a denominator in common, they get along really well, and we can do things like add them together. For instance:

$$\frac{1}{8} + \frac{2}{8} = \frac{3}{8}$$

We can just add the two numerators together and make a new fraction with the same denominator. You can also think of them as belonging to the same pizza, because $\frac{1}{8}$, $\frac{2}{8}$, and $\frac{3}{8}$ all come from a pizza that was cut into 8 total slices.

$$\frac{1}{8} + \frac{2}{8} = \frac{3}{8}:$$

Pretty easy, right? Subtraction works the same way. So as long as two fractions have a **common denominator**, you can just subtract across the top. For example:

$$\frac{7}{12} - \frac{2}{12} = \frac{5}{12}:$$

What's It Called?

Common Denominator

Two fractions share a *common denominator* when they have the same denominator (bottom); in other words, they have their denominators "in common."

Watch Out! Remember: it's the denominator (bottom) that has to be the same when you add and subtract fractions, not the numerator (top). And be sure that you keep the denominator the same in your answer, too—never add or subtract across the bottom.

 Doing the Math

Add or subtract these fractions together, but only if they have the same denominator. If they don't, just write "different denominator." (We'll learn how to handle those soon, too.) Always leave your answer in reduced form. I'll do the first one for you.

1. $\frac{6}{7} - \frac{1}{7} - \frac{5}{7}$

2. $1\frac{1}{3} + \frac{4}{3} =$ (Hint: make the mixed number an improper fraction first.)

3. $\frac{1}{3} + \frac{1}{7} =$

4. $\frac{7}{8} - \frac{3}{8} =$

5. $\frac{10}{9} + \frac{8}{9} =$

(Answers on p. 289)

Different Denominators

Often, you'll be asked to add or subtract two fractions with *different* denominators, like these two:

$$\frac{3}{5} + \frac{1}{2}$$

But do you see how these two guys don't get along? After all, they don't have the important "deep down" denominator in common.

And this means they *cannot* be added together until they *get* a common denominator. Consider the following example.

Look at this pepperoni pizza: one of its slices represents $\frac{1}{6}$.

And one of this veggie pizza's slices represents $\frac{1}{3}$.

But how could we possibly add:

What would we even write down? This is what would happen if we tried to add across the top and bottom. (Don't try this at home—it doesn't work!)

$$\frac{1}{6} + \frac{1}{3} \neq \frac{2}{9} \text{ slices}$$

Yikes! That's clearly not the way to go. Let's rethink this. Perhaps if we'd taken that veggie pizza and cut it up even more:

Now we can see how, instead of expressing the veggie pizza amount as $\frac{1}{3}$, we could express $\frac{1}{3}$ as two smaller slices: $\frac{2}{6}$. And we can add:

And we see how $\frac{1}{6} + \frac{1}{3} = \frac{1}{6} + \frac{2}{6} = \frac{3}{6}$. Now let's put our answer in reduced form: $\frac{3}{6} = \frac{3 \div 3}{6 \div 3} = \frac{1}{2}$. And voilà! $\frac{1}{6} + \frac{1}{3} = \frac{1}{2}$. We just added two fractions together by finding their common denominator.

Lowest Common Denominator (LCD)

Back on page 29, we reviewed how to find the lowest common multiple (LCM) of two numbers. (The LCM is the *smallest* multiple that two numbers share in common. The LCM of 4 and 3, for example, is 12.) When those two numbers happen to be *denominators of fractions*, then we call the LCM the **lowest common denominator** (**LCD**). The reason we care about lowest common denominators is that we can use them to add and subtract fractions that don't start out having the same denominator.

The LCM of 3 and 6 is 6. So in the pepperoni/veggie pizza example above, we rewrote the fractions so they both used the LCD of 6.

Remember, the *only* way to add and subtract fractions is if they have the same denominator. So what do you do if they don't have the same denominator? Use copycat fractions!

Step-By-Step

Adding/subtracting fractions with different denominators:

Step 1. Find the *lowest common multiple* of the two denominators—this is also called finding the LCD of the two fractions.

Step 2. Use copycat fractions to rewrite the fractions so they have the *same* denominator.

Step 3. Add the new fractions together.

Step 4. Reduce the final fraction, if possible.

And...
Action! Step-By-Step in Action

Here's how it works: let's say you want to add $\frac{1}{4} + \frac{3}{5}$.

(By the way, with a little practice, adding fractions like this won't take very long at all. But for now I'm going to explain every little step so you understand it. That's going to make it look longer than it really is, but at least you'll understand what the heck is going on, y'know?)

$$\tfrac{1}{4} + \tfrac{3}{5} = ?$$

We know we can't just add them together because they have different denominators, so what do we do? Here are the steps:

Step 1. Find the LCD. Well, the two denominators are 4 and 5, right? And we know from page 31 that the lowest common multiple of 4 and 5 is 20. (If we didn't already know that, then we would write out some of the multiples of 4 and 5 until we found the smallest one they have in common, or we could do the Birthday Cake method for finding LCMs, like on p. 30.)

Step 2. Use copycat fractions to rewrite the fractions so they have "20" as a denominator. If we want to rewrite $\tfrac{1}{4}$ so that it has a "20" on the bottom, then we ask the question: What do I have to multiply times 4 in order to get 20? We know from our multiplication tables that the number we're looking for is **5**, because $4 \times \mathbf{5} = 20$. We want the denominator to be 20, but we don't want to change the value of $\tfrac{1}{4}$, so we multiply by the copycat fraction $\tfrac{\mathbf{5}}{\mathbf{5}}$.

We know that the value of every copycat fraction is 1, and you can always multiply a fraction (or anything for that matter) by 1 without changing its value, so:

$$\tfrac{1}{4} = \tfrac{1}{4} \times \tfrac{\mathbf{5}}{\mathbf{5}} = \tfrac{1 \times \mathbf{5}}{4 \times \mathbf{5}} = \tfrac{5}{20}$$

Now let's do the same thing with our second fraction, which is $\tfrac{3}{5}$. We want to rewrite it so that it has a denominator of 20. So we ask the question: "What do I have to multiply times 5 in order to get 20?" And we know the answer is **4**, since $\mathbf{4} \times 5 = 20$. So, we'll use the copycat fraction $\tfrac{\mathbf{4}}{\mathbf{4}}$ to rewrite $\tfrac{3}{5}$ so that it has a 20 as its denominator.

$$\tfrac{3}{5} = \tfrac{3}{5} \times \tfrac{\mathbf{4}}{\mathbf{4}} = \tfrac{3 \times \mathbf{4}}{5 \times \mathbf{4}} = \tfrac{12}{20}$$

Step 3. Now that we've used our copycats, we can add our two new fractions together. So:

$$\tfrac{1}{4} + \tfrac{3}{5} = ?$$

becomes

$$\tfrac{5}{20} + \tfrac{12}{20} = \tfrac{17}{20}$$

which means:

$$\frac{1}{4} + \frac{3}{5} = \frac{17}{20}$$

Voilà!

QUICK NOTE! Sometimes the fractions you have to add or subtract can be reduced right off the bat. And if you try to do the problem with fractions that aren't reduced, you'll get stuck dealing with extra-big numbers, which can get messy! So always reduce your fractions before you start. It just makes life easier.

Take Two! Another Example!

What if you wanted to subtract $\frac{24}{32} - \frac{1}{12}$?

First off, let's reduce $\frac{24}{32}$, since we can, and it'll make the numbers smaller, which is always nice. We know that **8** divides into 24 and 32, so let's do $\frac{24}{32} = \frac{24 \div 8}{32 \div 8} = \frac{3}{4}$. Our problem now looks like this: $\frac{3}{4} - \frac{1}{12}$. That looks much nicer to deal with, and of course will give us the same final answer, because we haven't changed the *value* of the fraction we reduced.

Step 1. Find the LCD. The LCD is just the LCM of the two denominators, so let's list their first few multiples. (We also could have used the Birthday Cake method.)

4: 4, 8, <u>12</u>, 16
12: <u>12</u>, 24, 36

Okay, 12 is the LCD, since it's the lowest multiple they have in common.

Step 2. Rewrite the fractions with the LCD, 12, as the denominator, using copycats. Well, $\frac{3}{4}$ is the only one we need to change. We need to get 12 on the bottom, so let's multiply by the copycat $\frac{3}{3}$.

$$\frac{3}{4} = \frac{3}{4} \times \frac{3}{3} = \frac{3 \times 3}{4 \times 3} = \frac{9}{12}$$

Step 3. Now that they have the same denominator, subtract them: $\frac{9}{12} - \frac{1}{12} = \frac{8}{12}$

Step 4. Reduce, if possible. Yep, 8 and 12 both have **4** as a factor, so let's divide by **4** on top and bottom.

$$\frac{8 \div \mathbf{4}}{12 \div \mathbf{4}} = \frac{2}{3}$$

And there's our answer! $\frac{3}{4} - \frac{1}{12} = \frac{2}{3}$

Doing the Math

Add or subtract these fractions, by finding the LCD and rewriting the fractions using common denominators. I'll do the first one for you. (Always leave your answer in reduced terms!)

1. $\frac{1}{10} + \frac{2}{15} =$

<u>Working out the solution</u>: Let's find the LCD by listing the denominators' first few multiples (or you could use the Birthday Cake method discussed on p. 30).

$$\textbf{10: } 10, 20, \underline{30}, 40 \ldots$$

$$\textbf{15: } 15, \underline{30}, 45 \ldots$$

Now I'll rewrite both fractions so that they have 30 as their denominator, by using copycat fractions.

$$\tfrac{1}{10} = \tfrac{1 \times \mathbf{3}}{10 \times \mathbf{3}} = \tfrac{3}{30} \text{ and } \tfrac{2}{15} = \tfrac{2 \times \mathbf{2}}{15 \times \mathbf{2}} = \tfrac{4}{30}$$

Now add them together: $\frac{3}{30} + \frac{4}{30} = \frac{7}{30}$.

Since 7 and 30 don't have any common factors, the fraction is already reduced.

Answer: $\frac{7}{30}$

2. $\frac{7}{15} - \frac{1}{45} =$

3. $\frac{4}{9} - \frac{5}{12} =$

4. $\frac{1}{8} + \frac{2}{9} =$

5. $\frac{6}{18} + \frac{250}{300} =$ (Hint: reduce both fractions first!)

(Answers on p. 289)

QUICK NOTE! If you're asked to add or subtract two fractions with different denominators but you're feeling kind of lazy, the truth is, you don't actually have to find the LCD. Instead, you can just multiply the two denominators together and use *that* for a common denominator. For example, we added $\frac{1}{10} + \frac{2}{15}$ on p. 92; instead of starting by finding the LCD, 30, we could have been lazy and just used the common denominator $10 \times 15 = 150$. Then we'd use copycats to rewrite the fractions so they end up with denominators of 150: $\frac{1}{10} \times \frac{15}{15} = \frac{1 \times 15}{10 \times 15} = \frac{15}{150}$ and $\frac{2}{15} \times \frac{10}{10} = \frac{2 \times 10}{15 \times 10} = \frac{20}{150}$. Now we can add them: $\frac{15}{150} + \frac{20}{150} = \frac{35}{150}$, and we'd have to reduce: $\frac{35 \div 5}{150 \div 5} = \frac{7}{30}$. You'll often get bigger numbers and have to reduce like we just did, but sometimes it's faster in the long run—you can be the judge!

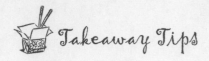 *Takeaway Tips*

- When two fractions have the *same denominator*, they are very compatible—they have a lot of the "deep down" stuff in common. So we say they have a *common denominator*.

- Fractions can only be added or subtracted to/from each other when they have a common denominator. Otherwise, they simply won't work with each other. And really, why should they? They have nothing important in common, after all!

- To add and subtract fractions that don't have a common denominator, we find the LCM of the denominators—in other words, the lowest common denominator (LCD)—and rewrite each fraction so that they *do* have a common denominator. Again, we're not changing the fractions' values when we do this; we're simply expressing the same values in different ways. Like recutting a pizza. *Mmm.*

Brain Push-Ups

The human brain is like a muscle, and just like your abs or biceps, it needs to be exercised regularly to stay in good shape.

You probably do some form of exercise, whether you play on a sports team or just go to gym class. You might not always feel like getting exercise, but you try to work it in because you know that it's good for you, right?

Plus, maybe you're close to your goal of how many push-ups you can do, and you know that if you keep practicing them, you'll keep increasing the number that you can do in a row. And that feels good!

Just like practicing push-ups, the more you practice math, the more you'll improve—and the smarter you'll feel.

Chapter 9

Choosing the Perfect Necklace

Complex Fractions

$\mathcal{S}$ay you're trying on an outfit for a party. You've got the dress, the shoes, and the earrings—and now you're choosing the right necklace. It's a wraparound dress, so you'll probably want a necklace that makes a V shape to mimic the neckline of the dress, right? That's probably going to be a necklace with a charm on a chain of some sort.

When you sort through your jewelry box, you spot a necklace that looks perfect—but it's all tangled! It's got a few small snarls and a huge knot in the middle—ugh. Before you can even try it on and see if it works, you'll have to untangle it. In other words, there's a "cleanup" step you must take care of before you can *do* anything with the necklace.

Just like the necklace above, **complex fractions** start off looking like a big mess—and before we can *do* anything with them, we have to "untangle" them.

What's It Called?

Complex Fractions

Complex fractions are fractions whose numerators and denominators *themselves* are expressed using fractions. Examples of complex fractions are $\dfrac{\frac{1}{3}}{\frac{5}{2}}$ and $\dfrac{\frac{1}{2} - \frac{1}{5}}{\frac{1}{5} + 2\frac{1}{2}}$.

My goal in this chapter is to make sure that no matter what kind of scary-looking complex fraction is thrown your way, you will be able to handle it with confidence and a clear head. Let's do it!

The simplest type of *complex fraction* is one in which the numerator and denominator are single fractions themselves.

$$\dfrac{\frac{1}{6}}{\frac{3}{4}}$$

In this example, the "numerator" of our complex fraction is $\frac{1}{6}$, and the "denominator" is $\frac{3}{4}$. But what does this *even mean*, and how can we simplify it? This sure doesn't look like a good ol' pizza fraction to me. (See p. 38 for good ol' pizza fractions.)

Remember when we talked about how fractions are *division problems* in disguise (see p. 42)? For example, the fraction $\frac{10}{5}$ "means" that we should divide the top by the bottom, $10 \div 5$, and we'd get 2. That is, $\frac{10}{5} = 2$. Well, that's one way to think about *complex fractions*, too.

To find the value of $\dfrac{\frac{1}{6}}{\frac{3}{4}}$, for example, we can simply divide the top by the bottom, which would be $\frac{1}{6} \div \frac{3}{4}$. You may recall that for fraction division (see p. 56), we simply flip the second fraction and multiply:

$$\frac{1}{6} \div \frac{3}{4} = \frac{1}{6} \times \frac{4}{3} = \frac{1 \times 4}{6 \times 3} = \frac{4}{18} = \text{(reducing)} = \frac{2}{9}$$

$$\text{so, } \frac{\frac{1}{6}}{\frac{3}{4}} = \frac{2}{9}$$

That's the basic method: fractions whose numerators and denominators are also fractions can be thought of as a big *division* problem, and can be simplified using regular fraction division. But now I'm going to show you a shortcut!

Shortcut Alert!
"MEANS AND EXTREMES"*

Whenever you have a fraction where the numerator and denominator are each fractions themselves, you can use the Means and Extremes method of simplifying: The very top and very bottom numbers are called the "extremes," and the numbers in the middle are called the "means." To simplify this complex fraction, all you have to do is *multiply the extremes together* (that gives you the numerator), and *multiply the means together* (that gives you the denominator).

$$\text{means} \left[\frac{\frac{1}{4}}{\frac{3}{5}} \right] \text{extremes} = \frac{1 \times 5}{4 \times 3} = \frac{5}{12}$$

I want to emphasize that in the Means and Extremes method, we are doing the *same things* to the numbers that we did in the "division" method I showed you above. The reason this method is faster is because we skip the steps of writing it out as a division problem and flipping the second fraction. It all happens at once!

Tall and Skinny

Ever seen a photo of a celebrity in which the celebrity looks *extremely* tall and skinny? Like, *too* tall and skinny? (Like, somebody *feed her* something!?) When looking at a photo like this, you might have

* FYI, the term means and extremes *is also sometimes used to describe a similar technique for solving proportions in algebra.*

thought to yourself, "*Extreme* measures need to be taken here. I *mean*, someone should make her eat something—*extremely* soon!"

Before you can apply the Means and Extremes method to a complex fraction, you have to make sure the fraction looks *extremely* "tall and skinny." That means it can't include any mixed numbers (which make it wider) or whole numbers (which make it shorter). It must always have 4 numbers stacked on top of each other—and that's it. The good news is, if you end up with mixed numbers or whole numbers, it's not hard to convert them into improper fractions (as we learned on p. 45).

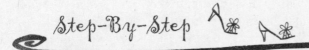

Not ready for Means and Extremes: $\dfrac{6\frac{1}{2}}{\frac{4}{78}}$ or $\dfrac{1\frac{1}{2}}{2}$

Ready for Means and Extremes: $\dfrac{\frac{13}{2}}{\frac{4}{78}}$ or $\dfrac{\frac{3}{2}}{\frac{2}{1}}$

Step-By-Step

The Means and Extremes shortcut (for simplifying complex fractions whose numerators and denominators are simple or improper fractions):

Step 1. Make sure that both the numerator and denominator are "tall and skinny"—either simple fractions or improper fractions (no whole numbers or mixed numbers).

Step 2. Multiply the very top and very bottom numbers (the extremes) to get the new numerator.

Step 3. Multiply the middle two numbers (the means) to get the new denominator.

Step 4. Reduce, and you're done!

Watch Out! Make sure that you always multiply the *extremes* for the new numerator and the *means* for the new denominator. How will you ever remember which to use? I'm glad you asked.

How NOT to Confuse the Means and Extremes

Here's how I've always kept them straight: Think about these tall skinny fractions: The very top and very bottom (the extremes) are more *exposed* than the middle numbers (means), which are more *sheltered* in the inside.

The numbers that are cozy in the middle are the ones that multiply together to create the *denominator* in the final fraction — they're accustomed to being more protected, so it's only natural that they'll still be sheltered by having a little "roof" over them in the final fraction, right? And the extremes are used to being more exposed to things like wind and rain (just go with me on this, okay?), so it's cool with them to still be "exposed" on the top of the final fraction as the *numerator*. Cozy stays cozy, and exposed stays exposed!

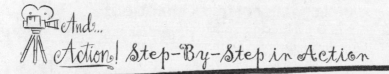

Let's do that first problem again, this time using the Means and Extremes method.

$$\frac{\frac{1}{6}}{\frac{3}{4}}$$

Step 1. Multiply the "exposed" extremes to get the new numerator.

Step 2. Multiply the "sheltered" means to get the new denominator.

$$\frac{\frac{1}{6}}{\frac{3}{4}} = \frac{1 \times 4}{6 \times 3} = \frac{4}{18}$$

Step 3. And now just reduce: $\frac{4}{18} = \frac{4 \div 2}{18 \div 2} = \frac{2}{9}$.

So, $\dfrac{\frac{1}{6}}{\frac{3}{4}} = \frac{2}{9}$.

As you can see, we got the same answer using the Means and Extremes method as we did using the division method (and in less time, too)!

QUICK NOTE! It's funny that an expression as complicated as $\dfrac{\frac{1}{6}}{\frac{3}{4}}$ could equal something as simple as $\frac{2}{9}$, but it's true. Remember that when you see a "=" sign, it means that the expressions on either side of the equation are equal to each other. They have the *same value*.

✂ *Shortcut Alert!*
CANCELING IN COMPLEX FRACTIONS
Try to cancel factors (a.k.a., reduce your fractions) right *after* the Means and Extremes part, but *before* multiplying anything out. This'll keep your numbers nice and small, which will make your life easier; you can eliminate big numbers before they happen. (Why does that sound like a bug spray advertisement?)

We first set up the Means and Extremes method:

$$\frac{\frac{24}{6}}{\frac{8}{3}} = \frac{24 \times 3}{6 \times 8}$$

But we don't multiply anything together yet. Otherwise we'd get big numbers. First, we cancel an 8 from the 24 and the 8, and then a 3 from one of the 3s and from the 6.

$$\frac{24 \times 3}{6 \times 8} = \frac{24 \times 3}{6 \times 8} = \frac{3 \times 3}{6 \times 1} = \frac{3 \times 3}{6 \times 1} = \frac{1 \times 3}{2 \times 1} = \frac{3}{2}$$

Doing the Math

Simplify these complex fractions using the Means and Extremes method. I'll do the first one for you.

1. $\dfrac{4}{3\frac{1}{3}} =$

<u>Working out the solution</u>: First, we can't have any whole numbers or mixed numbers when using the Means and Extremes method, so we'll rewrite 4 as $\frac{4}{1}$ and use the MAD Face method on $3\frac{1}{3}$ to rewrite it as the improper fraction $\frac{10}{3}$. So, now our complex fraction is ready for the "means and extremes."

$$\frac{\frac{4}{1}}{\frac{10}{3}} = \frac{4 \times 3}{1 \times 10}$$

We can cancel a 2 from the 4 and 10, and get:

$$\frac{\overset{2}{4} \times 3}{1 \times \underset{5}{10}} = \frac{2 \times 3}{1 \times 5} = \frac{6}{5}$$

It can't be reduced anymore, so that's our final answer!

$$\frac{4}{3\frac{1}{3}} = \frac{6}{5}$$

2. $\dfrac{\frac{5}{2}}{\frac{10}{12}} =$

3. $\dfrac{1\frac{1}{2}}{3} =$

4. $\dfrac{\frac{9}{28}}{\frac{3}{7}} =$

(Answers on p. 289)

QUICK NOTE! You can *only* cancel factors from the top and bottom of a fraction when the *only* operation in the numerator and denominator is *multiplication*. You cannot cancel factors if there is addition, subtraction, or division involved. For example, in $\frac{3+3}{6+1}$, you cannot cancel anything. If you tried to "cancel," you would <u>change the value</u> of the fraction, and that would be bad. Very bad.

Big, Scary Complex Fractions

Dealing with the "means and extremes" of a fraction is sort of like dealing with a small tangle or two in your necklace chain. But what if you had a big, scary-looking knot like this?

$$\frac{\frac{1}{4} + \frac{1}{2}}{2 - \frac{1}{8}}$$

How would you deal with it? Well, when you're untangling a piece of jewelry, you always do the easiest part first and work from the outside in. See the knot on the left?

You're better off dealing with the leftmost knot first and then perhaps the rightmost one. If you tried to tackle the whole big mess all at once, you'd be in trouble. We'll use the same strategy to "untangle" the big, scary complex fraction above.

Step-By-Step

Simplifying big, scary complex fractions:

Step 1. Calculate and simplify the numerator and denominator *separately*, until they are each simple fractions or improper fractions—never leave them as mixed numbers or whole numbers.

Step 2. Do the Means and Extremes method to make the whole thing a simple or improper fraction: multiply the sheltered inside numbers (the means) to get the denominator and the exposed outside numbers (the extremes) to get the numerator.

Step 3. Reduce. (Maybe you were even canceling along the way according to the "Shortcut Alert!" on p. 100.)

And... Action! Step-By-Step in Action

Let's untangle $\dfrac{\frac{1}{4} + \frac{1}{2}}{2 - \frac{1}{8}}$.

Step 1. Calculate/simplify the numerator and denominator separately. First we'll look at the numerator and pretend that nothing else in the world exists. Here's the numerator: $\frac{1}{4} + \frac{1}{2}$. So, what does this numerator equal? Let's add the two fractions together. (See p. 89 for a review of adding fractions with different denominators.) Their LCD will be 4, so we can rewrite $\frac{1}{2}$ as $\frac{1 \times 2}{2 \times 2} = \frac{2}{4}$.

So the numerator becomes: $\frac{1}{4} + \frac{2}{4} = \frac{3}{4}$. Good. Now we can rewrite our big, scary complex fraction as: $\dfrac{\frac{3}{4}}{2 - \frac{1}{8}}$. Hmm. It's still a little scary.

Okay, next knot! Let's tackle the denominator:

Looking only at the denominator and pretending that nothing else exists, we need to simplify $2 - \frac{1}{8}$. We know how to do this; we just need to express 2 as an improper fraction and then do fraction subtraction (see p. 89 for a review of how to subtract fractions with different denominators). So let's rewrite it as: $\frac{2}{1} - \frac{1}{8}$. The LCD will be 8, so we rewrite $\frac{2}{1}$ as $\frac{2 \times 8}{1 \times 8} = \frac{16}{8}$, and now we can subtract them: $\frac{2}{1} - \frac{1}{8} = \frac{16}{8} - \frac{1}{8} = \frac{15}{8}$.

So the denominator of our big, scary fraction now looks like $\frac{15}{8}$.

And our big, scary fraction is now just a regular *complex fraction*: $\dfrac{\frac{3}{4}}{\frac{15}{8}}$.

And now we're left with a knot that we're used to untangling. Time to use the Means and Extremes method! Let's do it, using our canceling method to avoid big numbers (see p.100). We can cancel a 3 from the 3 and the 15, and then we can cancel a 4 from the 4 and the 8:

$$\frac{3 \times 8}{4 \times 15} = \frac{\overset{1}{\cancel{3}} \times 8}{4 \times \underset{5}{\cancel{15}}} = \frac{1 \times 8}{4 \times 5} = \frac{1 \times \overset{2}{\cancel{8}}}{\underset{1}{\cancel{4}} \times 5} = \frac{1 \times 2}{1 \times 5} = \frac{2}{5}$$

Yep, after all that untangling, that big, scary fraction ended up being equal to sweet little $\frac{2}{5}$.

Now that we've untangled all the knots, what do you think? Does your necklace make the outfit?

Order of Operations Review

Because the numerators and denominators of big, scary fractions can get complicated, this is a good time to review the "PEMDAS" **order of operations**...and the dining habits of Pandas.

Ah, pandas. Aren't they cute? And they have really big appetites. I've heard that pandas like to eat dumplings with mustard, and then for dessert, they have apples with spice—like cinnamon or nutmeg. Yum!*

....

* *The truth? Pandas eat mostly bamboo. But they do come from China, and Chinese dumplings are yummy, especially with hot mustard. Plus, pandas eat apples in the zoo, so I'm not WAY off. Besides, this'll help you remember how to use the Order of Operations correctly, and that's sort of more to the point, isn't it?*

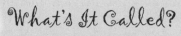

What's It Called?

Order of Operations

The *Order of Operations* is the order in which we must simplify math expressions:

P̲ANDAS	P̲arentheses
E̲AT	E̲xponents
M̲USTARD ON D̲UMPLINGS	M̲ultiplication & D̲ivision
-AND-	
A̲PPLES WITH S̲PICE	A̲ddition & S̲ubtraction

"P̲andas E̲at M̲ustard on D̲umplings and A̲pples with S̲pice!"

I recommend saying this out loud a few times, until you've learned it—you'll see that it's got a rhythm to it. Notice that the panda eats two separate courses: dinner and *then* dessert. Multiplication and division happen together at dinner; they have the *same priority*, and we do whichever one comes first, left to right. (So I could have said "Dumplings with Mustard" instead of "Mustard on Dumplings.")

The same is true for addition and subtraction (Apples & Spice): They happen together during dessert, so they have the same priority as each other, and we just do whichever one (addition or subtraction) comes *first*, left to right. PEMDAS might as well have been called PEDMAS or even PEDMSA—see what I mean?

Let's use the panda rule to correctly simplify the expression:

$$12 - 8 \div 2 + 5 \times (2 + 1) = ?$$

We first simplify inside the P̲arentheses (P̲andas!), and since $2 + 1 = 3$, we get:

$$12 - 8 \div 2 + 5 \times \mathbf{3} = ?$$

The next word is "E̲at," but since there are no E̲xponents,* we move on to the first panda meal: M̲ustard on D̲umplings. It's time to find all the M̲ultiplication and D̲ivision, and do them, left to right. D̲ivision comes first in this example, so since $8 \div 2 = 4$, we get: $12 - \mathbf{4} + 5 \times 3 = ?$

Next we do the M̲ultiplication, and since $5 \times 3 = 15$, we get: $12 - 4 + \mathbf{15} = ?$ Now all that's left is dessert! So we'll do the A̲ddition

••••••••••••••••••••

* *If you don't know what exponents are yet, that's okay—just skip this part for now!*

and Subtraction (Apples & Spice!), left to right. In this example, subtraction comes before addition. Since 12 − 4 = 8, we get: **8** + 15 = ? And finally, we can add 8 + 15 = **23**.

So the answer is: $12 - 8 \div 2 + 5 \times (2 + 1) = 23$

Now that you've gotten a refresher on the *order of operations*, let's go back to dealing with our big, scary complex fractions.

 Take Two! Another Example!

Sometimes complex fractions get even scarier—but now that we know the key is to simply untangle the numerator and denominator separately, then proceed as we would with any *complex fraction*, we can handle anything, right?

$$\frac{\frac{5}{12} + 3 \div \frac{1}{2}}{3\frac{1}{2} \times \left(\frac{1}{4} - \frac{1}{8}\right)} = ?$$

Uh, yeah. Don't panic—we'll get through this together.

Just as before, we will tackle the numerator and denominator *separately*. First, let's look at the numerator only, as if nothing else in the world exists:

$$\frac{5}{12} + 3 \div \frac{1}{2}$$

We may be tempted to add $\frac{5}{12} + 3$, but we should *always* follow the order of operations, which says that division comes before addition. So the first thing we'll do is divide $3 \div \frac{1}{2}$.

As you know, in fraction division, we flip the second fraction and multiply across. Also, you should always write whole numbers as *improper fractions* before you do "fraction operations" with them.

$$3 \div \frac{1}{2} = \frac{3}{1} \times \frac{2}{1} = \frac{3 \times 2}{1 \times 1} = \frac{6}{1}$$

Now the numerator looks like: $\frac{5}{12} + \frac{6}{1}$. If we wanna add these guys together, we need a common denominator, right? Certainly, the LCM of 12 and 1 is 12, so let's use that and rewrite $\frac{6}{1}$ to have 12 as its denominator.

$$\frac{5}{12} + \frac{6}{1} = \frac{5}{12} + \frac{6 \times \mathbf{12}}{1 \times \mathbf{12}} = \frac{5}{12} + \frac{72}{12} = \frac{77}{12}$$

So the numerator of the big scary fraction is $\frac{77}{12}$. Not so bad!

Let's move on to the big scary denominator.

$$3\tfrac{1}{2} \times \left(\tfrac{1}{4} - \tfrac{1}{8}\right)$$

Order of operations tells us to deal with whatever's inside the parentheses first. That's $\tfrac{1}{4} - \tfrac{1}{8}$. Their LCD will be 8, so let's make sure both fractions are written with a denominator of 8:

$$\tfrac{1}{4} - \tfrac{1}{8} = \tfrac{1 \times \mathbf{2}}{4 \times \mathbf{2}} - \tfrac{1}{8} = \tfrac{2}{8} - \tfrac{1}{8} = \tfrac{1}{8}$$

Alright, now the big scary *denominator* looks like: $3\tfrac{1}{2} \times \tfrac{1}{8}$. We know better than to attempt fraction multiplication without first converting the mixed number into an improper fraction, so that's what we'll do, using the MAD Face method.

$$3\tfrac{1}{2} = \tfrac{7}{2}$$

So now the big scary *denominator* looks like: $\tfrac{7}{2} \times \tfrac{1}{8}$. And multiplying two fractions isn't so bad (see p. 51). We now get $\tfrac{7}{2} \times \tfrac{1}{8} = \tfrac{7 \times 1}{2 \times 8} = \tfrac{7}{16}$.

And finally, combining our big scary numerator with its big, scary denominator, we get the following "tall and skinny" complex fraction:

$$\frac{\frac{5}{12} + 3 \div \frac{1}{2}}{3\tfrac{1}{2} \times \left(\tfrac{1}{4} - \tfrac{1}{8}\right)} = \frac{\frac{77}{12}}{\frac{7}{16}}$$

This looks much more manageable. Let's do the "means and extremes" to get the final answer.

$$\frac{\frac{77}{12}}{\frac{7}{16}} = \frac{77 \times 16}{12 \times 7} \text{ (We'll cancel a 7 and then a 4.)}$$

$$\frac{\overset{11}{\cancel{77}} \times 16}{12 \times \underset{1}{\cancel{7}}} = \frac{11 \times 16}{12 \times 1} = \frac{11 \times \overset{4}{\cancel{16}}}{\underset{3}{\cancel{12}} \times 1} = \frac{11 \times 4}{3 \times 1} = \frac{44}{3}$$

It can't be reduced any further, so we're done untangling (*pant, pant*).

So that must mean: $\dfrac{\frac{5}{12} + 3 \div \frac{1}{2}}{3\tfrac{1}{2} \times \left(\tfrac{1}{4} - \tfrac{1}{8}\right)} = \tfrac{44}{3}$. Done!

QUICK NOTE! Let's say you were asked to do something with these big, scary complex fractions, like add two of them together. No worries! The key is simply to deal with the big, scary complex fractions *one at a time* and to simplify each of them separately until they are simple or improper fractions. *Then* add them together. It all comes down to untangling that necklace, undoing the outside knots first, and patiently making your way to the center.

Doing the Math

Simplify these expressions. I'll do the first one for you.

1. $\dfrac{\frac{1}{4} + \frac{1}{2}}{2 - \frac{1}{8}} + \dfrac{\frac{5}{12} + 3 \div \frac{1}{2}}{3\frac{1}{2} \times \left(\frac{1}{4} - \frac{1}{8}\right)} =$

<u>Working out the solution</u>: We should deal with each fraction separately, and *then* think about adding them together. Since we've already simplified each of these complex fractions in this chapter (see p. 103 and p. 106), we know this problem actually simplifies down to:

$$\tfrac{2}{5} + \tfrac{44}{3} = ?$$

The LCD of 5 and 3 is 15, so let's rewrite both fractions so they both have 15 on the bottom—then we'll be able to add them: $\frac{2 \times 3}{5 \times 3} + \frac{44 \times 5}{3 \times 5} = \frac{6}{15} + \frac{220}{15} = \frac{226}{15}$. Is this reduced? The only factors of 15 are 3 and 5. We know that 226 isn't divisible by 5 because it doesn't end in a 5 or 0. Let's add up the digits to see if they are divisible by 3. (See the EZ divisibility tricks chart on p. 9 to review the 3s rule.)

Since neither 3 nor 5 divides into 226, and those are the only factors of 15, we know our fraction is reduced.

<u>Answer</u>: $\frac{226}{15}$

2. $\dfrac{\frac{2}{5} - \frac{1}{3}}{\frac{1}{30}} + 2 =$

3. $\dfrac{\frac{3}{4} + \frac{1}{4} \div \frac{1}{3}}{\frac{1}{2} - \frac{1}{6}}$ = (Hint: Remember the order of operations!)

4. $\dfrac{2\frac{1}{5} + \frac{1}{2}}{\frac{5}{4} \div \frac{1}{9} - \frac{9}{4}} + \frac{1}{5}$ =

(Answers on p. 289)

What's the Deal?

If you have any brain power left, you may be wondering (because of the "order of operations" rule) why we could *add* stuff in the numerator before we even looked at the *multiplication* in the denominator. That's an excellent question! And here's the answer: the numerator and denominator of big, scary fractions really *are* in their own separate worlds—the best way to think of numerators and denominators is as if they each have their own set of <u>parentheses</u> around them. As you know, you always *completely* simplify whatever's in the parentheses first, before moving on—and that's what we did!

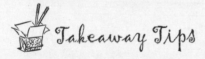 Takeaway Tips

- When you see a big, scary complex fraction, *untangle the numerator and the denominator separately.* Just think of it as untangling knots in your necklace chain—with a little patience and dedication, it'll all unravel just fine!

- *Before you can use* the Means and Extremes method, a complex fraction should look "tall and skinny." No mixed numbers or whole numbers, just two simple or improper fractions stacked on top of each other.

- When working with complex fractions, you can wait until the end to cancel, but if you cancel factors *before* you multiply the numbers out (right after "means and extremes") it will help to keep your numbers much smaller along the way.

- Always remember the Order of Operations: Parentheses, Exponents, Multiplication & Division (whichever comes first), and Addition & Subtraction (whichever comes first). "Pandas Eat Mustard on Dumplings and Apples with Spice!" And treat the numerators and denominators of big, scary fractions as if they have parentheses around them.

Quiz #2:
Do You Have Trouble Focusing—
Or Are You a "Focus Superstar"?

Take this quiz and find out!

Do you know how to sit down and focus or are you a bit of a scatterbrain? Let's see what expert psychologist Robyn Landow, PhD, says!

1. You've just sat down to do homework, when your girlfriend IMs you, saying she has some really good gossip. You:

   a. Chat for a minute or so but then let her know that you are in the middle of something and will get back to her later.

   b. Suddenly realize you left IM on by mistake! Oops. You quickly type, "Let's talk later," and sign off. You know that if you don't study, you won't be ready for your test!

   c. Intend to make it a quick response, but before you know it, it's been an hour!

2. Complete this sentence: *When I am speaking with someone, I would describe my focus as . . .*

   a. Okay. But I may "tune out" from time to time if something is on my mind.

   b. Solid. I look them in the eye and really listen.

   c. Not so great. I tend to think more about what I am going to say back to them or what I am going to do next.

3. How would you describe your study habits?

   a. "Slow and steady wins the race."

b. "A little bit of this, a little bit of that."

c. "Cramming," with a few late-nighters thrown in.

4. It's Monday, and your teacher just announced that this Friday's test will cover the last three topic areas. You:

a. Hold off on studying until Thursday. After all, you want to study as close to the test date as you can.

b. Take a couple of nights off and start studying on Wednesday.

c. Plan out the next four days, assigning a topic to study each night and leaving room for an overall review on Thursday.

5. Complete this sentence: *When I do my homework, I . . .*

a. Turn off my cell phone and close my email and IM. I don't like any distractions!

b. Talk on the phone, listen to music, IM and/or surf the web at the same time. It keeps my mind alert—otherwise I get bored.

c. Listen to music, take the occasional phone call, and often answer an IM for a few minutes—but just for a few minutes.

6. On nights when you have an extreme amount of homework, you:

a. Keep working until you get everything finished. Once you stop, you want it to be over with.

b. Set reasonable blocks of study time and reward yourself with appropriate breaks.

c. Keep feeling a sense of dread at the prospect of all that work—so you try to cheer yourself up by putting it off and doing something more fun first!

7. You usually write about 10 items on your "to do" list each day. By the end of the day, you have completed:

a. Almost all of them.

b. No more than half of them—the ones you haven't completed get transferred to tomorrow's list.

c. "To do" list? What "to do" list?

8. Your friend just called and spilled her guts about a bad situation she'd gotten herself into after school. She's fine now—she just needed to vent. What do you remember most about the conversation 10 minutes later?

a. You had some really great advice for her—so mostly you remember your advice.

b. Um . . . she felt better after spilling her guts, right? But since you were also occasionally answering IMs on "silent mode," you didn't catch some of the details.

c. There were some things she revealed about herself that you made mental notes of, since they might help you to help her avoid similar situations in the future.

9. It's bedtime. You:

a. Get ready, go to bed, and sleep soundly 'cause you've got tomorrow covered.

b. Get ready, go to bed, then write yourself a note about something you forgot to do, so you'll remember to do it in the morning.

c. Freak out when you realize there are at least three things you forgot to do for tomorrow!

10. Describe your locker at school.

a. Yikes! I need to do a *major* clean out! There are old homework assignments, a couple of jackets, and probably some overdue library books shoved in there.

b. I can always find exactly what I need. I line up my books so they don't get squished. I also put a little magnetic mirror inside so I can check my hair between classes.

c. I manage to fit all my stuff in my locker, but sometimes I wish I kept things a little neater.

11. How would your friends describe your organization skills?

   a. My friends would say I'm good about getting all my homework done, but they're surprised at how often I forget the little details.

   b. My friends always want to know how I keep track of everything so well. In my group of friends, I'm definitely the one who stays focused and gets things done.

   c. My friends know not to put me in charge of details or planning.

12. When math homework gets particularly difficult, you:

   a. Call your friend and ask her what answer she got. At least then you will know what you need to end up with.

   b. Skip the assignment and plan to ask someone about it the next day at school—but let's face it, that may or may not happen.

   c. Go back to your notebook to review the notes from the day, or refer to an outside book like *Math Doesn't Suck* for help. If you're still getting tripped up by something, you may call a friend to discuss where you got stuck and see if she can remind you what the next step is.

Scoring:

1. a. 2;	b. 3;	c. 1	**7.** a. 3;	b. 2;	c. 1
2. a. 2;	b. 3;	c. 1	**8.** a. 2;	b. 1;	c. 3
3. a. 3;	b. 2;	c. 1	**9.** a. 3;	b. 2;	c. 1
4. a. 1;	b. 2;	c. 3	**10.** a. 1;	b. 3;	c. 2
5. a. 3;	b. 1;	c. 2	**11.** a. 2;	b. 3;	c. 1
6. a. 2;	b. 3;	c. 1	**12.** a. 2;	b. 1;	c. 3

30–36 points
Congratulations—you have already developed some excellent study habits and are naturally good at focusing! You know to stay on top of your assignments, and you reach out to friends or teachers for help when you need it. But don't forget to give yourself a rest as well! If you are tired, angry, distracted, or in a hurry, your brain will not retain the information you are studying. Keep nourishing your brain and body with water, healthy food, exercise, and sleep. What you put in is what you will get out!

20–29 points

You are pretty good about getting things done, but you slack off from time to time. Do you have your priorities straight? How is your scheduling? Generally, if you designate certain times of the day for studying, you will get into a routine and accomplish more. If you just "fit it in" during your day, chances are that there will never be enough time. Put your study time down on your calendar—kind of like a doctor's appointment. For example: "Tuesday 3:00–4:30, STUDY." And be sure to make this "appointment" at a time when your energy and focus level is at its highest. You may think you are a night owl, but most students are more alert during the daytime and early evening hours. And if you postpone chatting, surfing, texting, and so on until after your goals for that study session are over, you will be able to give your friends your undivided attention! Create rewards for yourself for successfully completing a task, such as calling a friend or watching that show you recorded or downloaded.

12–19 points

Having a bit of trouble, huh? Can't seem to focus or get your work done in an organized way? A good rule of thumb is that studying should be done when you are rested, alert, and have planned for it. Last-minute studying just before bed or just before a class is usually a waste of time. Once you make a schedule for studying, don't be afraid to change it. Schedules are meant to reflect how you truly intend to use your time, not how you think you should! If it doesn't work, change it. (See the paragraph above for more schedule-related tips.) Remember, avoiding studying is the easiest thing in the world. While doing homework, "pay attention to your attention." When you find yourself having difficulty concentrating, switch to a different subject or take a short break. Studies show that you start forgetting information soon after you hear it—so, especially if you're having trouble focusing on your teacher in class, review the material ASAP when you get home to secure the information in your memory. Also, ask yourself questions as you study. That way, you know you are not just running your eyes over the paper. If you like music in the background as you study, that's fine, but don't let it be a distraction. And avoid your cell phone, telephone, IM, etc. while you study—this will help you concentrate and will demonstrate to those around you that this is not a time to be disturbed!

Chapter 10

What Every Savvy Shopper Should Know

All About Decimals

Reality Math

*Y*ou want to buy a magazine that costs $2.70, but there is a 7% tax that gets added at the register. If you only have $3 on you, will you be able to buy it?

Being able to work easily with decimals—and to convert and compare them with fractions and percents—is *so* important. Every gal who likes to shop should know how this stuff works. Oh yeah, and it comes up in homework, too. So this chapter is filled with tips to make decimals easier!

Decimals

What Are They?

Here's what a decimal number looks like:

tens
ones
tenths
thousandths
hundreds
decimal point
hundredths

153.275

And 153.275 can be expanded into its "parts":

$$153.275 = 153 + 0.2 + 0.07 + 0.005$$

Because when you add up 153 + 0.2 + 0.07 + 0.005, you get 153.275!

QUICK NOTE! Adding Zeros

Zeros all the way to the right, *after* the decimal, have no effect on a number's value. For example:

2.4 has the *same value as* **2.40**, in other words, **2.4 = 2.40**

0.05 has the *same value as* **0.050**, in other words, **0.05 = 0.050**

427 has the *same value as* **427.00000**, in other words, **427 = 427.00000**

This might seem silly or obvious, but it comes in handy.

Also, for some reason, we always put a "0" before the decimal for numbers less than 1. Honestly, I think the only reason we do this is so the little decimal isn't mistaken for a piece of dirt on the paper.

.14   0.14

dirt?   oh... it's a decimal.

Watch Out! You can't just add zeros *anywhere* you want, though. For example, 2.4 does *not* equal 2.04. If you add a zero *after* the decimal place, it has to be *all* the way at the end: 2.4 = 2.40 = 2.400 and so on.

Comparing Decimals

Sometimes it's easy to compare two decimals, like 0.2 < 0.5, because 2 is smaller than 5. (Remember, the "alligator mouth" always wants to eat the larger number!) But what about 0.0098 and 0.021? Or 0.45099 and 0.45106?

Imagine there are two girls who just finished a gymnastics routine in an important meet, and they are now being judged by several judges. Who will win, Leslie or Robin?

Just like the digits in a decimal, the judges are sitting in order of importance from left to right— the one on the left is the most important, and the one on the right is the least. In fact, the judges become so much less important as you move to the right that their scores only count if the first judges' scores are tied!

So, let's say the first judge gives both girls a "4."

Then the next judge gives them both a "5." Hmm, still a tie, so let's move on to the third judge.

The third judge gives Leslie a "0" and Robin a "1." I guess she's kind of a tough judge. But since Robin got a bigger number than Leslie, we have a winner! It doesn't matter what the fourth judge thinks, because as soon as one judge gives a different score, *all the judges to the right don't matter.*

So, we could conclude that 0.**450**99 < 0.**451**06 and never even have to look past the third decimal place!

When you're comparing two decimals, think of our judges and, starting from left to right, use your hand (or a little, imaginary hand) to cover up all of the digits except the *first* one after the decimal point—because they won't matter unless there is a tie. So, to compare 0.0098 and 0.021, we'd do:

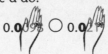

0.0098 ○ 0.021

So far we are 0 to 0—a tie. Let's move our hand over and check the next digit.

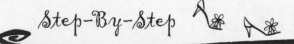

$$0.00\cancel{98} \bigcirc 0.02\cancel{1}$$

Well, a "2" beats a "0," and because it's no longer a tie, we can ignore the rest of the digits and declare the winner! 0.021 is bigger than 0.0098. In other words:

$$0.0098 < 0.021$$

QUICK NOTE! When comparing decimals, if there are numbers to the left of the decimal place, like in 29.06 and 3.25, always look at those *whole numbers* first, and if they are different, well, use your common sense to tell you that 29.06 is bigger than 3.25. (After all, $29.06 is greater than $3.25, right?) Obviously the whole numbers "count" for a lot more than their decimal companions!

Step-By-Step

Comparing decimals:

Step 1. If there are whole numbers to the *left* of the decimal point, and they are *different*, you already know which number is bigger (the one with the larger whole number). If the whole numbers are the same or they don't exist, continue.

Step 2. Cover up all the digits to the *right* of the decimal place except the first digit. Compare *just that one digit* in each number. If one of them is bigger, then you've found which number is bigger. If it's a tie, move your hand to the right one place on both numbers and continue to compare *one digit at a time* until you find the side that is bigger.

Step 3. Once one side has a bigger digit, then you can *ignore* the rest of the digits after them, and you're done!

Doing the Math

Compare these decimals—which one is bigger? I'll do the first one for you.

1. 9.0688 ◯ 9.62

Working out the solution: To the left of the decimal point, we have two 9s, so the whole numbers are equal. Now let's move on to the decimal part. At first, we might be tempted to call the one on the left bigger, because we see a 688 and a 62, but going digit by digit, and covering up everything to the right of the first digit after the decimal point, we see that 0 < 6. Since there is no longer a tie, we have a winner and can ignore the rest of the digits.

Answer: 9.0688 < 9.62.

2. 0.8888 ◯ 0.891

3. 0.45 ◯ 0.1999

4. 56.11 ◯ 6.889

5. 0.1112 ◯ 0.1211

(Answers on p. 289)

Adding and Subtracting Decimals

Adding and subtracting decimals isn't too different from the regular addition and subtraction you're used to. The trick is to make sure that the decimal points are directly lined up with each other (which I'll show you how to do below). After that, it's a breeze!

Step-By-Step

Adding and subtracting decimals:

Step 1. Arrange the numbers so that the <u>decimal points line up directly</u> on top of each other. You can add zeros to the end of the numbers if that helps with lining them up.

Step 2. Then just add or subtract as usual, and be sure to pull the decimal point *directly down* to the answer, too. Done!

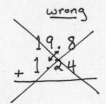

And... Action! Step-By-Step in Action

Let's add: 19.8 + 1.24

Steps 1 and **2.** Arrange the numbers so the decimals line up over each other. Make sure the *decimal points* line up over each other. (I like putting in extra zeros when needed—see below—because it helps me to keep my columns straight while I'm adding. And when subtracting, those extra zeros make it much easier to "regroup"—a.k.a. "borrow"— when you need to!)

wrong
~~19.8~~
~~+ 1.24~~

right
$$19.80$$
$$+ 1.24$$
$$\overline{2\,1.04}$$

← I added this zero to make the columns easier for me to see.

Answer: 21.04

Take Two: Another Example

With subtraction, all the same regrouping rules apply. Seriously, after you line up your decimal points correctly (and stick in your extra zeros where needed), you can pretty much forget that there are even decimals involved!

Let's subtract: 6.01 − 0.791

Step 1. Arrange the numbers so the decimal points line up, adding zeros if you want.

Step 2. Subtract normally, and make sure to put the decimal directly below in the answer.

$$
\begin{array}{r}
{}^{5}\cancel{6}.{}^{9}\cancel{0}{}^{10}\cancel{0}0 \\
- \ \ 0\ !\ 7\ 9\ 1 \\
\hline
5\ !\ 2\ 1\ 9
\end{array}
$$

← I added this zero so it would be easier to regroup, and to see the columns.

Answer: 5.219

 Doing the Math

Add or subtract, and remember to line up your decimal points! I'll do the first one for you.

1. 7 − 0.09

<u>Working out the solution</u>:

$$
\begin{array}{r}
{}^{6}\cancel{7}.{}^{9}\cancel{0}\cancel{0} \\
- \ \ 0\ !\ 0\ 9 \\
\hline
6\ !\ 9\ 1
\end{array}
$$

→ I added these zeros so it would be easier to regroup.

<u>Answer</u>: 6.91

2. 3.001 + 21.4

3. 0.59 + 73.001

4. 6.11 − 0.5

5. 32 − 4.5

(Answers on p. 289)

Multiplying Decimals

Multiplying Decimals by 10

The number 10 is one of the easiest numbers to multiply by, don't you think? After all, $3 \times 10 = 30$, $7 \times 10 = 70$, and so on. It was one of the easiest times table columns for me to learn, that's for sure.

Well, decimals are all based on the number 10, and <u>moving a decimal place to the right or left is the same as multiplying or dividing by 10</u>. Check it out:

Getting bigger...

$0.032 \times 10 = 0.32$; $0.32 \times 10 = 3.2$; $3.2 \times 10 = 32$; $32 \times 10 = 320$

Getting smaller...

$320 \div 10 = 32$; $32 \div 10 = 3.2$; $3.2 \div 10 = 0.32$; $0.32 \div 10 = 0.032$

So, if you had a number like 24.59 and someone said, "*Multiply* this by 10!" all you'd have to do is *move the decimal point* to the *right* one place so that the number gets bigger: 245.9.

And if someone instead said, "*Divide* this number by 10," all you'd have to do is *move the decimal point* to the *left* one place so that the number gets smaller: 2.459.

Easy, huh?

When would you multiply decimals in your life? Anytime you're dealing with money!

Reality Math

Multiplying Decimals by Numbers Other than Ten

Say you're standing in line at the local sandwich place, picking up lunch for yourself and 4 friends. Sandwiches are $4.75 each, including tax. You have $24.13 with you. Do you have enough money, or should you run home and ask Mom for more cash before getting all the way up to the front of the line?

To figure out how much money you'd need, you could just grab a napkin from the counter, fish out a pen from your fabulous bag, and multiply 4.75×5 (since there are 4 friends + you, which makes 5 people).

But how do you multiply decimals? No problem.

It's just like regular multiplication, except, at the end, you have to **count** how many total numbers came after the decimal in the two numbers you started with, then apply that to your final answer. Here's how it works. For $4.**75** × 5, you would count the **two** numbers after the decimal point, then *drop* the decimals, and multiply: 475 × 5 = 2375.

$$2\,3.\underset{\frown}{75}$$

You then put the **two** decimal places back in: $23.**75.** And that's it!

$23.75? Yep, you have enough money. No need to run home and get more.

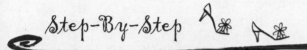

Step-By-Step

Multiplying decimals: count, multiply, count:

Step 1. Count the total number of digits to the right of the decimal points in both numbers.

Step 2. Drop the decimal points and **multiply** both numbers together.

Step 3. Starting from the right, **count** the same number of total decimal places back, and insert the decimal point. Done!

And... Action! Step-By-Step in Action

Multiply 0.45 × 11.3

In this example, *both* numbers are decimals.

Step 1. *Count* the total digits after the decimal points:

0.**45** × 11.**3**

That's a total of **three** decimal places.

Step 2. Then, drop the decimals and *multiply* the numbers: 45 × 113 = 5085.

Step 3. Finally, since we got **three** for the total number of digits after the decimal points, starting from the right, *count* **three** places back, and so the answer is 5.085. Done!

More About Zeros

When you are starting a problem and you see a number like 1.80, you know that you can drop that final zero, since 1.8 = 1.80. But does that zero *matter* when you're counting decimal places in decimal multiplication?

That's a good question. And here's the answer: *As long as you are consistent with your zeros and the decimal places, it'll all work out.*

Let's try an example both ways: 1.80 × 0.4

If you drop the zero to begin with, 1.8 × 0.4, here's how it would go: 1.**8** × 0.**4**: count the **two** numbers after the decimal points, multiply 18 × 4, get 72, and then put the **two** decimal places back in to get the answer, 0.72.

And if you leave the zero in there, 1.80 × 0.4, here's how it would go: 1.**80** × 0.**4**: count the **three** numbers after the decimal points, multiply 180 × 4, get 720, and then put the **three** decimal places back in to get the answer, 0.720.

As you can see, since 0.720 = 0.72, both ways give us the same answer!

You've Seen Her on TV!

"$\mathcal{I}$ think that everyone has the ability to be smart, but it takes a strong-willed, focused mind to use intelligence in positive ways. I have a deep respect for girls—and all people—who choose to educate themselves.

"Now that I live on my own, math means bills, balancing bank accounts, and overall money management—and knowing how to handle these things makes me feel successful—and 'grown up!'" Valery Ortiz, Madison Duarte on the N's *South of Nowhere*

Doing the Math

Multiply these decimals. I'll do the first one for you.

1. $0.40 \times 8.3 =$

Working out the solution: Notice that $0.40 = 0.4$, so let's drop that zero to keep things simple. Now we're going to **count, multiply, count**: 0.4×8.3. We **count** that it has **two** numbers total after the decimal points, so drop the decimal points and **multiply** $4 \times 83 = 332$. Now put the decimal point back in, **counting two** places back, to get 3.32.

Answer: 3.32

2. $0.60 \times 0.30 =$

3. $9.1 \times 1.00 =$

4. How about the question from the beginning of the chapter: You want to buy a magazine that costs $2.70, but there is a 7% tax that gets added at the register. If you only have $3 on you, will you be able to buy it? (Hint: 7% = 0.07)

(Answers on p. 290)

Decimal Division

First, I'll show you how to divide *into* decimals. These problems have *whole numbers* on the outside of the division house and *decimals* inside, like this: $3\overline{)4.2}$. These kinds of decimal division problems are pretty easy.

Second, I'll show you how to divide *by* a decimal, so that the number on the outside is a decimal, too, like $0.3\overline{)4.2}$ or $0.3\overline{)42}$. These aren't so hard; they just require an extra step, which I'll show you.

First, let's do a quick review of some division words.

What's It Called?

Divisor, Quotient, Dividend

Here are some tips for remembering the different parts of a division problem (because from time to time you'll have to know which is which, and it can be hard to remember the difference between a divisor and a dividend!).

$$\text{divisor} \rightarrow 3\overline{)12} \leftarrow \text{dividend}$$
$$4 \leftarrow \text{quotient}$$

The little $\overline{)}$ thingy is sometimes called a "division box" or "division bracket" . . . but I like to think of it as a little "division house."

One day in the fifth grade, when I was feeling really frustrated because I couldn't remember which part of the problem was the **divisor** and which was the **dividend**, I created this little picture inside my head—maybe it'll help you, too.

The dog has his *eye* on the steak inside the house, so he's called the "div-EYE-sor" (which is how you pronounce *divisor*). And his *end* goal is to get inside to that yummy steak, so the number inside is the "divid-END" (*dividend*).

Then for the **quotient**, imagine that there is a little person standing on the roof, talking to the dog, and I *quote*, "Sparky, you'll get your dinner later—stop staring!" This might help you to remember that the *quotient* goes where the person we are *quoting* would be standing—on the roof of the house.

Of course, division problems are often written in a different form, such as $12 \div 3 = 4$ or $\frac{12}{3} = 4$. If you're asked to identify the divisor, dividend, or quotient when a problem is written like this, just translate the problem into the form that uses the $\overline{)}$, imagine the little dog and the steak, and you'll have the answer right in front of you!

 ## Doing the Math

In each problem, label the divisor, dividend, and quotient. I'll do the first one.

1. $\frac{6}{2} = 3$

Working out the solution:
First let's rewrite it in "house" form, $2\overline{)6}$ with quotient 3

Answer: Divisor = 2, Dividend = 6, and Quotient = 3.

2. $3\overline{)63}$ with quotient 21

3. $32 \div 4 = 8$

4. $\frac{80}{5} = 16$

5. $72 \div 8 = 9$

6. $\frac{10}{2} = 5$

(Answers on p. 290)

Dividing into Decimals

Dividing *into* decimals happens all the time in real life, because that's what we do when we divide up money! For example: Let's say you are going to the arcade with 4 friends, and among you, you have a total of $32.50—which you want to turn into quarters as soon as you can. If you divide the money evenly, how much will each of you get?

Well, you'd need to divide 5 *into* $32.50, in other words: $5\overline{)32.50}$.

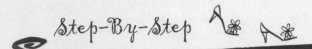

Dividing into decimals:

Step 1. Pull the decimal up through the roof of the little dividing house thingy.

Step 2. Divide normally. Done!

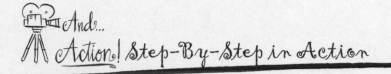

Going back to the arcade game problem:

$$5\overline{)32.50}$$

Steps 1 and **2.** Pull the decimal directly above, onto the roof, and proceed with regular division.

$$5\overline{)32\overset{.}{.}50} \;\rightarrow\; \begin{array}{r} 6.50 \\ 5\overline{)32.50} \\ -30 \downarrow \\ \hline 25 \\ -25 \\ \hline 00 \end{array} \rightarrow \text{DONE!}$$

So you'd each get $6.50. And how many *quarters* is that? Since there are 4 quarters in every dollar, you could multiply $6.50 by 4: 6.5 × 4 = 26. That makes 26 quarters each. Nice!

Take Two! Another Example!

So, if you divide $3\overline{)4.2}$, you'll get 1.4, and that makes sense because 4.2 is only a little bit bigger than 3, so our answer should be just a little bit bigger than 1.

In decimal division, you can divide a bigger number into a *smaller* number, and it works exactly the same way. What if the number inside the dividing house (the dividend) were 10 times smaller, like $3\overline{)0.42}$? Or 100 times smaller, like $3\overline{)0.042}$? Let's see how they compare:

$$
\begin{array}{r}
1.4 \\
3\overline{)4.2} \\
-3\downarrow \\
\hline
12 \\
-12 \\
\hline
0 \rightarrow \text{DONE!}
\end{array}
\quad \text{vs.} \quad
\begin{array}{r}
0.14 \\
3\overline{)0.42} \\
-3\downarrow \\
\hline
12 \\
-12 \\
\hline
0 \rightarrow \text{DONE!}
\end{array}
\quad \text{vs.} \quad
\begin{array}{r}
0.014 \\
3\overline{)0.042} \\
-3\downarrow \\
\hline
12 \\
-12 \\
\hline
0 \rightarrow \text{DONE!}
\end{array}
$$

For this last one, we get a really, really small number. Remember, the decimal place always has to go *directly* above on the roof, and sometimes that means that we need to fill in spaces with zeros on the roof, like we just did in this example. Since the 1 had to go *directly* above the 4, we needed to fill in that spot in between the decimal point and the 1 with a zero. No problem!

QUICK NOTE! Adding More Zeros in the Middle of Division

Here's something kinda nice about decimal division. Since you can always add zeros to the right of a decimal without changing the value of the number, like

$$0.31 = 0.310 = 0.3100 = 0.31000, \text{ and so on,}$$

if you are dividing something like $2\overline{)0.31}$, and you need more zeros, you can add them!

$$
\begin{array}{r}
0.15 \\
2\overline{)0.31} \\
-2\downarrow \\
\hline
11 \\
-10 \\
\hline
1 \leftarrow \text{hmm... not done yet!}
\end{array}
\qquad
\begin{array}{r}
0.155 \\
2\overline{)0.310} \leftarrow \text{add zeros if you need them!}\\
-2\downarrow \\
\hline
11 \\
-10\downarrow \\
\hline
10 \\
-10 \\
\hline
0 \rightarrow \text{DONE!}
\end{array}
$$

Incidentally, you can also use this trick when you don't have decimals in the division problem, but you get something like 2)$\overline{43}$. As you may already know, you can always add a decimal point after the 43 and add a zero, because 43 = 43.0. They have the *same value*. So if you find that you can't finish the long division problem without resorting to a fraction or a remainder (and you'd rather have a decimal answer than a fraction answer), then just add the decimal and zero on the fly and continue solving the problem . . .

$$
\begin{array}{r}
21.5 \\
2\overline{)43.0} \\
-4\downarrow \\
\hline
03\downarrow \\
-2\downarrow \\
\hline
10 \\
-10 \\
\hline
0 \rightarrow \text{DONE!}
\end{array}
$$

. . . and voilà!

 ## Doing the Math

Solve these division problems, adding zeros and even decimal points if you need them (as long as you don't change the value of the numbers in doing so). I'll do the first one for you.

1. 6)$\overline{0.15}$

Working out the solution:

$$
\begin{array}{r}
0.025 \\
6\overline{)0.150} \quad \leftarrow \text{add zeros if you} \\
-12\downarrow \quad \text{need them!} \\
\hline
30 \\
-30 \\
\hline
0 \rightarrow \text{DONE!}
\end{array}
$$

2. 4)$\overline{0.52}$

3. 4)$\overline{0.052}$

4. 5)$\overline{67}$

(Answers on p. 290)

Dividing by Decimals

So far, we've dealt only with problems where we want to divide *into* a decimal number. Now we'll see what happens when we divide *by a* decimal number: This is when the number on the *outside* of the little dividing house is a decimal. For example:

$$26.5\overline{)848}$$

But how would this ever be useful in real life? Tons of ways!

Reality Math

Say you're burning a DVD of a bunch of home-video minimovies that you recorded with your digital camera. They're really cute. You've been documenting every week of your new puppy's life with a minimovie dedicated to him. He's 9 months old now, so you have a lot of weeks stored up!

Each movie takes up about 26.5 MB of space on the DVD. You've also burned a bunch of puppy *pictures* onto the DVD, so there's only about 848 MB of free space left. How many of these 26.5 MB mini-movies do you have room for on the DVD?

To find the answer, we'd need to see how many times 26.5 MB can go into 848 MB, right? In other words, we'd need to solve $26.5\overline{)848}$.

But wait—how do we divide when a decimal point is in the divisor? (If you don't have a calculator handy, that is.)

Here's the trick: in long division, the divisor *can't* have any numbers after the decimal point, so we <u>move the divisor's decimal place</u> <u>to the right as many times as we need to in order to get a whole number</u>, and <u>then move the *dividend's* decimal place that many times, too</u>. Here's the step-by-step.

Step-By-Step

Dividing by decimals:

Step 1. "Get rid" of the decimal in the divisor: Move the decimal as many places to the right as it takes to make the divisor (the number you're dividing by) a whole number. **Count** how many places you moved the decimal.

Step 2. Then *also* move the decimal point that **same number** of places in the dividend (the number you're dividing into, inside the little dividing house). Note: sometimes you'll have to add zeros to make this happen!

Step 3. Divide normally. Done!

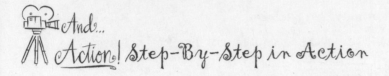

And...
Action! Step-By-Step in Action

Let's divide: 0.54 ÷ 0.09

And of course, this is the same as $0.09\overline{)0.54}$

Steps 1 and 2. First, let's "get rid of" the decimal in the divisor. For 0.09, we have to move the decimal point **two** places to the right to get a whole number, which means we also need to move the decimal **two** places in the dividend, 0.54:

$$0.09 \rightarrow 0.09_{\curvearrowright} = 9 \qquad 0.54 \rightarrow 0.54_{\curvearrowright} = 54$$

Now our new problem looks like this: $9\overline{)54}$

Step 3. Divide normally: $9\overline{)54}^{\,6}$ (If you remembered your times tables, this step was easy!)

Answer: 0.54 ÷ 0.09 = 6.
Not so bad, eh? Let's do another one.

Take Two! Another Example!

Back to our example with the puppy minimovies: $26.5\overline{)848}$

Step 1. First, we "get rid of" the decimal in the divisor. Let's see how many places we have to move it. Well, 26.5 has **one** number after the decimal point, so we'll move the decimal point **one** place to the *right* to get 265. We've successfully gotten rid of the decimal.

Step 2. Now, since we moved the decimal point **one** place to the right for 26.5, we have to move the decimal point **one** place to the right for the number 848. Since there is no decimal point in 848, however (it's a whole number), we have to remember that 848 is the same thing as 848.0, and *then* move the decimal point **one** place to the right, to get 8480. (You can also think of this step as adding zeros—whichever makes the most sense to you.) Our new problem is $265\overline{)8480}$

Step 3. Divide normally: $265\overline{)8480}$

$$
\begin{array}{r}
32 \\
265\overline{)8480} \\
-795\downarrow \\
\hline
530 \\
-530 \\
\hline
0 \rightarrow \text{DONE!}
\end{array}
$$

scratch work:

$$
\begin{array}{cc}
265 & 265 \\
\times\ 3 & \times\ 2 \\
\hline
795 & 530
\end{array}
$$

Answer: 32

So, we did our long division and found out that we can store 32 puppy minimovies on the DVD. Nice!

QUICK NOTE! Remember that moving the decimal point to the right or left is the same as multiplying or dividing by 10 or 100 or 1000, and so on. For example, if you multiplied 0.75 by 100, you'd get 75. And if you divided 0.75 by 100 you'd get 0.0075.

Doing the Math

Remember, decimal division is all about decimal points and zeros! I'll do the first one.

1. $2.1 \div 0.05 =$

<u>Working out the solution:</u> This is the same as $0.05\overline{)2.1}$. We need to move the decimal point two places in both

numbers, so the divisor won't have a decimal point in it: $5\overline{)210}$. And now just divide normally: $5\overline{)210}$ with answer 42 above.

<u>Answer</u>: 42

2. $0.8\overline{)4}$ =

3. $4.2 \div 0.6$ =

4. $4.2 \div 0.60$ =

5. $0.8\overline{)0.4}$ =

(Answers on p. 290)

What's the Deal?

Return of the Copycat

If you're wondering why we are "allowed" to move the decimals in decimal division, then you might like to read this section.

I'll show you that moving the decimals—a.k.a. multiplying *both the divisor and the dividend* by 10 or 100—is just like using a *copycat fraction!*

$$\text{divide } 1.2 \div 0.06$$

Typically, we'd now write this as $0.06\overline{)1.2}$ and move the decimal point two places in the divisor and dividend to get $6\overline{)120}$. But let's see why this is "allowed." Remember that you can always write division as a fraction, like this:

$$1.2 \div 0.06 \rightarrow \frac{1.2}{0.06}$$

But yikes! $\frac{1.2}{0.06}$ looks kind of scary, so let's get rid of those decimal places right away. Hmm. What would we have to multiply times the bottom to get rid of two decimal places? 100.

But we know if we want to multiply the bottom by 100 we also have to multiply the top by 100—to use a copycat fraction, in other words, because you can always multiply by a copycat without changing the value of your fraction. So:

$$\frac{1.2}{0.06} = \frac{1.2}{0.06} \times \frac{100}{100} = \frac{1.2 \times 100}{0.06 \times 100} = \frac{120}{6}$$

And guess what? Tip over $\frac{120}{6}$ for fraction division and you get $6\overline{)120}$ —exactly what we got by just moving the decimal point two places in the divisor and dividend!

So, multiplying by a copycat fraction that has 10 or 100 or 1000 on top and bottom is *essentially what we're doing* when we move the decimal point one, two, or three places in the divisor *and* dividend of a division problem. (And remember: When you multiply by a copycat, you don't change the value of the expression.)

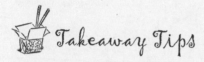 *Takeaway Tips*

- When *comparing decimals*, think of the gymnastics judges! Use your hand to cover up all the digits except the first one to the right of the decimal place, and then compare *just that digit*. If it's a tie, then move over one more place and repeat until it's no longer a tie. Once you have a winner, you can ignore the rest of the digits—you're done!

- When *multiplying decimals*, it's all about *counting* decimal places, using regular *multiplication* in between, and *counting* again to add the decimal back in: **count, multiply, count**.

- To tell apart the *divisor*, *dividend*, and *quotient*, think of the story about the dog *eye*-ing the steak (his *end* goal) in the house, and the little *quoting* girl on the roof. You'll never confuse them again!

- When *dividing <u>into</u> decimals*, just pull the decimal point of the dividend directly up to the roof of the little "dividing house," add zeros where you need them, and divide normally.

- When *dividing <u>by</u> decimals*, move the decimal point of the *divisor* to the right so it becomes a whole number, then move the decimal point *the same number of places to the right in the dividend*, and divide normally.

Chapter 11

Why Calculators Would Make Terrible Boyfriends

Converting Fractions and Mixed Numbers to Decimals

Reality Math

$\mathcal{S}$ay you have $50, and you want to buy a fabulous blue sundress that costs $62. Bummer! Not enough money. But wait, there's a sale tag that says it's $\frac{1}{5}$ off. Do you have enough money now?

Well, it says to take $\frac{1}{5}$ *off* the price, which means that if we find $\frac{4}{5}$ of $62, then we'll know the sale price of the dress, right?

This dress is way too cute to pass up, so let's find $\frac{4}{5}$ of $62. (Note that, when the word *of* is surrounded *immediately* by two numbers, then *of* means "multiply." More on this in chapter 15.)

$$\frac{4}{5} \text{ of } \$62 = \frac{4}{5} \times 62 = \frac{4}{5} \times \frac{62}{1} = \frac{4 \times 62}{5 \times 1} = \frac{248}{5}$$

Ugh. I mean, $\frac{248}{5}$ doesn't exactly look like an amount of money, does it? What's a shopping gal to do?

This is a perfect example of when to **convert a fraction to a decimal**.

Converting Fractions → Decimals

FRACTIONS ARE DIVISION PROBLEMS IN DISGUISE!

There's a factoid you may remember from chapter 4: fractions are *division problems* in disguise—and *dividing* is exactly how we're going to go from fractions to decimals. Seriously, if you can divide, you can convert fractions into decimals!

Consider the easy fraction $\frac{1}{2}$. You probably know that $\frac{1}{2}$ is the same as 0.5. Well, let's treat the fraction like the division problem it really is, by pushing the 1 over to the right*: $\frac{1}{2} \rightarrow 2\overline{)1}$

$$\frac{1}{2} = 2\overline{)1} = 2\overline{)1.0} = 2\overline{)\begin{smallmatrix} 0.5 \\ 1.0 \\ -1\ 0 \\ \hline 0 \rightarrow DONE! \end{smallmatrix}}$$

And look at that—we got 0.5. We just showed that $\frac{1}{2} = 0.5$, through division.

You may also know that $\frac{1}{4} = 0.25$. After all, a quarter is $\frac{1}{4}$ of a dollar. Let's try dividing 1 by 4 and see what happens—we should get 0.25, right?

$$\frac{1}{4} = 4\overline{)1} = 4\overline{)1.00} = 4\overline{)\begin{smallmatrix} 0.25 \\ 1.00 \\ -8\downarrow \\ \hline 20 \\ -20 \\ \hline 0 \rightarrow DONE! \end{smallmatrix}}$$

Yep. And remember how I told you that the value of every copycat fraction is 1? Let's demonstrate that with the copycat fraction $\frac{4}{4}$. We'll do the division problem, and see what it equals.

---

* *Remember from page 42, you can also read fractions from top to bottom: "1 divided by 2."*

$$\frac{4}{4} = 4\overline{)4} = \begin{array}{r} 1 \\ 4\overline{)4} \\ \underline{-4} \\ 0 \end{array} \rightarrow \text{DONE !}$$

That's got to be the easiest division problem I've ever seen. Yep, $\frac{4}{4} = 1$.

And that's it: to convert fractions to decimals, just divide!

QUICK (REMINDER) NOTE! Remember, *when dividing with decimals,* that when you divide a bigger number into a smaller number, you can always stick a decimal at the end of the smaller number and then add as many zeros as you need to, in order to finish the division. For example, $2\overline{)1}$ can become $2\overline{)1.0}$ or even $2\overline{)1.00}$. (Remember that adding zeros *after* a decimal place, all the way to the right, doesn't change the value of the number.)

Then just pull the decimal up through the top of the little "dividing house" and continue to divide like you normally would—the decimal on top just ends up being a part of the answer. And it doesn't matter if you add too many zeros—if you don't need 'em, don't use 'em. Their feelings won't get hurt, I promise!

Step-By-Step

Converting fractions → decimals:

Step 1. Do the division problem by tipping over the top: $\frac{top}{bottom} = bottom\overline{)top}$

Step 2. Add a decimal point and zeros in case you need them: $bottom\overline{)top.\mathbf{000}}$.

Step 3. Proceed with regular ol' decimal division, and you're done!

Let's work on converting fractions → decimals.

Remember the blue sundress from the beginning of the chapter? We needed to know if the $50 we had in our wallet was enough to pay for it. The dress was $\frac{1}{5}$ *off* of $62, so we calculated $\frac{4}{5}$ of $62 and got $\frac{248}{5}$.

But to know what $\frac{248}{5}$ means for our *wallets*, we need to convert $\frac{248}{5}$ into a decimal. I bet you know what to do now.

Step 1. Tip over the fraction and get the division problem: $5\overline{)248}$

Step 2. Divide, adding zeros where we need them and putting a decimal up top: $5\overline{)248.\overset{\bullet}{0}}$

Step 3. Proceed with regular decimal division!

$$
\begin{array}{r}
49.6 \\
5\overline{)248.0} \\
-20\downarrow \\
\hline
48 \\
-45\downarrow \\
\hline
30 \\
-30 \\
\hline
0 \rightarrow \text{DONE!}
\end{array}
$$

$49.6 \Rightarrow \$49.\underline{60}$

So, on sale, the sundress costs $49.60—and since we have $50, we can buy it! (Oh yeah, they were also having a "no tax" special, so there's no tax.) Being *math savvy* does come in handy sometimes.

Calculator Tip

Anytime you are allowed to use calculators, you can easily convert fractions to decimals by just turning them into division problems and dividing on the calculator.

So, if you needed to change some really weird fraction like $\frac{90616}{1928}$ into a decimal, all you'd have to do is think of it as a division problem, $1928\overline{)90616}$, which of course is $90616 \div 1928$, and plug it into your calculator. Just make sure you get the order right! An easy way to double-check that you've gotten it right is if you have an improper fraction like $\frac{90616}{1928}$, where the top is bigger than the bottom, then you know that the answer should be more than 1, because all improper fractions are greater than or equal to 1. And if you have a regular fraction like $\frac{2}{17}$, where the top is smaller than the bottom, then after you do the division in your calculator you should get an answer that is less than 1. No, this isn't where I talk about how calculators would make terrible boyfriends—we'll get to that later. . . .

Doing the Math

Convert these fractions into decimals. I'll do the first one for you.

1. $\frac{3}{8} =$

<u>Working out the solution</u>: Let's tip this fraction over and divide.

$$\frac{3}{8} \xrightarrow{\text{tip it over!}} \begin{array}{r} 0.375 \\ 8\overline{)3.000} \\ -24 \\ \hline 60 \\ -56 \\ \hline 40 \\ -40 \\ \hline 0 \end{array} \to \text{DONE!}$$

<u>Answer</u>: 0.375

2. $\frac{2}{5} =$

3. $\frac{1}{8} =$

4. $\frac{6}{4} =$

(Answers on p. 290)

"*This year, I entered a higher level math class, and suddenly it became challenging. As the class got harder, I found myself working harder and, if you can believe it, enjoying it. I have found that when something is beyond your reach, it makes you want it all the more.*

"*I think that math gives people a chance to feel like they are good at something, to feel proud of themselves. When you reach to understand something that is difficult, you learn that you can be smart, and possibly even be great at something.*" Maddie, 12

Converting Mixed Numbers → Decimals

Let's say that instead of a fraction, you have a mixed number like $6\frac{2}{5}$, and you want to turn *that* into a decimal. No problem!

QUICK NOTE! Remember that $6\frac{2}{5}$ is another way of writing $6 + \frac{2}{5}$. In other words, $6\frac{2}{5}$ equals 6 *whole* pizzas *plus* $\frac{2}{5}$ of another pizza. (See p. 40 for a review of mixed numbers.)

Step-By-Step

Converting mixed numbers → decimals:

You have two choices for how to tackle this. They both will work; it's simply a matter of your personal taste.

Method 1. You can *divide just the fraction part* and then *add* the resulting decimal number to the whole number.

Method 2. Or, you can *convert the whole thing into an improper fraction* and then divide.

When dividing, just follow the "steps" for converting fractions to decimals (see p. 139).

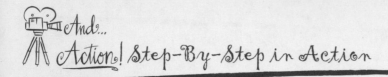
Convert $6\frac{2}{5}$ to a decimal.

Using Method 1: First consider that $6\frac{2}{5} = 6 + \frac{2}{5}$, and deal with the fraction part first. Tip over $\frac{2}{5}$ and divide $5\overline{)2.0}$ → $5\overline{)2.0}^{\,0.4}$. Now reunite both parts together: $6 + 0.4 = 6.4$

Using Method 2: First convert the entire thing to an improper fraction (use MAD Face technique from p. 45): $6\frac{2}{5} = \frac{32}{5}$. Next, divide 32 by 5. A little decimal division (see p. 125 for a review) will show you that: $5\overline{)32.0}^{\,6.4} = 5\overline{)32.0}$. And there you go—it's the same answer!

You should use whichever method you find easiest. Isn't it nice to have a *choice* in math every now and then?

 Doing the Math

Convert these mixed numbers into decimals. I'll do the first one for you.

1. $3\frac{4}{5} =$

<u>Working out the solution</u>: Remember that $3\frac{4}{5} = 3 + \frac{4}{5}$. I'll use method 1 from above. First, I'll convert the fraction part, $\frac{4}{5}$, into a decimal, by dividing the bottom into the top: $5\overline{)4.0}$ → $5\overline{)4.0}^{\,0.8}$. And then I'll add it to the whole number part, 3, to get our total answer.

<u>Answer</u>: $3 + 0.8 = 3.8$

2. $1\frac{1}{5} =$

3. $2\frac{3}{4} =$

4. $3\frac{1}{2} =$

(Answers on p. 290)

Converting Fractions ➔ Repeating Decimals

Sometimes when you are dividing a fraction to get its decimal form, the long division just keeps going and going . . . forever.

You might be familiar with $\frac{1}{3}$ = 0.3333333333 . . . The "..." means that the pattern of 3s keeps repeating—yes, forever. Another way to say that a decimal number keeps repeating itself is to put a little bar over the repeating part: $\frac{1}{3} = 0.\overline{3}$.

Let's do the division and see how these repeating decimals come to be. We'll tip the fraction over and divide, adding zeros where we need them: $\frac{1}{3} \rightarrow 3\overline{)1.0}$.

$$\frac{1}{3} = \quad 3\overline{)1.000} \quad \begin{array}{l} 0.333 \\ -9\downarrow \\ \overline{10} \\ -9\downarrow \\ \overline{10} \\ -9\downarrow \\ \overline{10} \\ -9 \end{array}$$

Yikes! You can tell that this will *never* stop, no matter how many zeros we add.

At this point, we throw our arms up and say, "Fine, be that way. I'm just going to put a little bar over your head and be done with it!" And that's how it comes to be that $\frac{1}{3} = 0.\overline{3}$.

Step-By-Step

Converting fractions to repeating decimals:

Step 1. Do the division problem by tipping over the top: $\frac{top}{bottom} = $ bottom$\overline{)top}$

Step 2. Add **zeros** in case you need them, and put the decimal point above: bottom$\overline{)top.\overset{\bullet}{000}}$

Step 3. If you notice a *repeating pattern* in the decimals that just keeps going and going, put a $\overline{\text{bar}}$ over the repeating part, and you're done!

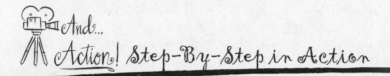

What's It Called?

Repeating Decimal

A *repeating decimal* (also called a *recurring decimal*) is a decimal expression that has a repeating pattern after the decimal point. For example, $0.33333\ldots$, $8.818181\ldots$, and $0.123123123123\ldots$ are all repeating decimals. These could also be written using bars, like this: $0.\overline{3}$, $8.\overline{81}$, and $0.\overline{123}$.

And... Action! Step-By-Step in Action

Let's convert these fractions (whose decimal form will be repeating).

$$\frac{92}{66} = ?$$

Hey, before we start, can we reduce this? Yep—because they are both even. Let's do that, to make our division a little easier.

$$\frac{92}{66} = \frac{92 \div 2}{66 \div 2} = \frac{46}{33}$$

Any more factors they have in common? Well, 33 is only divisible by 3 and 11, and neither of those numbers divides evenly into 46, so I guess we've found its simplest form. Fine, let's move on to the steps.

Steps 1 and **2.** Tip $\frac{46}{33}$ over and set up the division, adding zeros and the decimal.

$$\frac{46}{33} =$$

```
        1 . 3 9 3 9
   33 ) 46 . 0 0 0 0
       - 33 ↓
         1 3 0
       -  9 9 ↓
         3 1 0
       - 2 9 7 ↓
         1 3 0
       -  9 9 ↓
         3 1 0
       - 2 9 7
```

pattern keeps repeating

scratch work:

$$\begin{array}{r} 33 \\ \times\ 3 \\ \hline 99 \end{array} \qquad \begin{array}{r} {}^{2}33 \\ \times\ 9 \\ \hline 297 \end{array}$$

Notice that 39 keeps repeating over and over, so our answer is $\frac{92}{66} = 1.\overline{39}$.

Watch Out! Always make sure the numbers after the decimal are *really repeating* before you stop and put a bar over them. For example, compare $\frac{2}{9}$ with $\frac{9}{40}$, and pay special attention to the *subtraction* part of the division problems:

$$\frac{2}{9} = 9\overline{)2.0000} \quad \begin{array}{r} 0.222\ldots \\ \end{array}$$

$$\text{subtraction}\atop\text{keeps}\atop\text{repeating} \left\{ \begin{array}{r} -18\downarrow \\ \hline 20 \\ -18\downarrow \\ \hline 20 \\ -18\downarrow \\ \hline 20 \end{array} \right.$$

$$\frac{9}{40} = 40\overline{)9.0000} \quad \begin{array}{r} 0.225 \\ \end{array}$$

$$\text{subtraction}\atop\text{does not}\atop\text{repeat} \left\{ \begin{array}{r} -80\downarrow \\ \hline 100 \\ -80\downarrow \\ \hline 200 \\ -200 \\ \hline 0 \rightarrow \text{DONE!} \end{array} \right.$$

So $\frac{2}{9} = 0.\overline{2}$; it is repeating. But if you stopped too soon, you can see how you might think $\frac{9}{40}$ was a repeating decimal, too. We get two 2s in a row, right? But in fact, $\frac{9}{40} = 0.225$, which doesn't repeat at all! So how do you know when it's safe to stop dividing and put a bar over a repeating decimal?

The best way to tell whether a decimal is repeating or not is whether or not the *subtraction* you are doing (as part of the long division) is repeating again and again.

Notice for $\frac{9}{40}$, how we are subtracting 80 from 90 and then 80 from 100—it keeps changing. Whereas for $\frac{2}{9}$, the actual subtraction repeats: we keep subtracting 18 from 20, over and over again.

Take Two! Another Example!

$$\frac{7}{12} = ?$$

Steps 1 and **2.** Divide 12 into 7 and add zeros and the decimal point.

$$12\overline{)7.00000} \quad \begin{array}{r} 0.5833 \\ \end{array}$$

$$\text{only this part}\atop\text{will keep}\atop\text{repeating} \left\{ \begin{array}{r} -60\downarrow \\ \hline 100 \\ -96\downarrow \\ \hline 40 \\ -36\downarrow \\ \hline 40 \\ -36\downarrow \\ \hline 40 \end{array} \right.$$

Answer: $\frac{7}{12} = 0.58\overline{3}$.

Notice that not *all* of the numbers after the decimal point will necessarily be repeating. So make sure you put the $\overline{\text{bar}}$ only over the part that repeats!

These *repeating decimals* happen all the time in math, and when you're converting fractions into decimals, you never know when they might pop up.

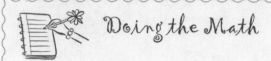

Doing the Math

Convert these fractions into decimals. Some will be repeating, some won't. I'll do the first one for you.

1. $\frac{5}{33} =$

<u>Working out the solution</u>: Since the fraction is already in its lowest, reduced terms, we just do the long division.

$$\frac{5}{33} = 33\overline{)5.00000} = 0.1515...$$

scratch work:

$$\begin{array}{r} 33 \\ \times\ 5 \\ \hline 165 \end{array}$$

<u>Answer</u>: $\frac{5}{33} = \mathbf{0.\overline{15}}$. Done!

2. $\frac{4}{15} =$

3. $\frac{6}{15} =$

4. $\frac{23}{33} =$

5. $1\frac{1}{9} =$ (Hint: remember that you can either convert this into an improper fraction first and divide, or you can choose to deal with the whole number and fraction parts separately, and then add them together after dividing the fraction part. Whichever method you prefer.)

(Answers on p. 291)

Watch Out! A word to the wise about calculators and repeating decimals: Calculators don't know how to handle the concept of *forever*. They only see things in the short term. (Sort of like some guys we all know.)

When a calculator sees a number that keeps going forever, it doesn't know what to do except round it off.

For example, a little long division will easily show that $\frac{2}{3}$ = 0.$\overline{6}$, but if you divide 2 ÷ 3 in a calculator, you'll get something like 0.666666666667. What's that 7 doing in there? The calculator doesn't understand that the 6s go on forever, and so it just rounds off the last digit it can display.

But 0.666666666667 would be the wrong answer.

Because they can't handle the idea of *forever*, calculators are bad at repeating decimals—and would make terrible boyfriends.

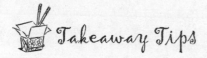 *Takeaway Tips*

- When converting a fraction to a decimal, just tip the fraction over and do long division. After all, fractions are just division problems in disguise.

- When converting a mixed number to a decimal, you have two choices. Either convert it to an *improper fraction* before converting to a decimal—or convert *just the fraction part* to a decimal and then add that to the whole part.

- If the division problem keeps going and going, and the decimals show a *repeating pattern*, then stop dividing and put a little $\overline{\text{bar}}$ over the repeating part.

- Always *reduce your fractions* before you start, if you can. It's nicer to deal with smaller numbers when it comes time to divide!

TESTIMONIAL:

 Stephanie Peterson (Dallas, TX)

Before: Struggling math student
Today: Petroleum analyst, actress,
web designer

I grew up in a very troubled home, and I spent most
of my evenings watching TV or listening to music—
anything that would keep my mind off of my parents
and their problems. Because I didn't work very hard,
whenever it came time for a math test, I would always
end up feeling anxious, regretful, and disappointed.
Often, when I got my exams back, I would cry because I
had gotten a C or a D.

> "Often, when I got my exams back, I would cry because I had gotten a C or a D."

Then, in college, I enrolled as a
theater major. (Theater was my first
love.) I only had to take one math class
to graduate, and I wanted to get it over
with as soon as possible. Freshman year,
I signed up for Math in Society, which I
hoped would be a joke of a class. Wrong!
Not only was the material interesting
(much to my surprise), but my professor
was this snazzy lady named Jill Dumesnil.
She was brilliant—and very girly!

A few weeks into the semester, Professor
Dumesnil noticed how hard I was working and
asked if I would like to tutor for the course. She said,
"You know, you're really good at this. You can do this."
No one had ever encouraged me in math like that before.

I said yes—and immediately regretted it. I hated
math, right? But once I started helping other people
discover that they could do math, too, I was hooked.
I fell deeper and deeper in love with mathematics
with every course I took. I ended up double-majoring
in math and acting!

Today I use my math skills every day in my job as a petroleum analyst. A *petrole-what?*

Basically, I work with engineers to analyze the economics of oil and gas. We give companies or individuals an estimate as to how much oil and gas they might get out of the land they own and how much money they might make from it.

What's so glamorous about all of this? Well, besides the fabulous job conditions (my own, beautiful office—yay!) and salary, I love what I do!

I've also learned that your life doesn't have to come down to just *one* job. I recently launched my own Web design business, www.ActingWebDesign.com. And what happened to acting? I'm still doing that, too. In fact, I met Danica McKellar on the set of one of her TV shows, *Inspector Mom*—that's when she asked me to share my story with all of you!

Chapter 12

How to Entertain Yourself while Babysitting a Devil Child

Converting Decimals to Fractions

$\mathcal{Y}$ou know how in the last chapter I said that if you can *divide* then you can change fractions into decimals? Well, get this: if you can *count*, you can change decimals into fractions. Seriously! It's not hard to do, and it's a handy skill in everyday life for all us gals.

Converting Decimals → Fractions

Converting decimals to fractions comes down to realizing that *decimals* can be rewritten as fractions with a denominator of 10 or 100 or 1,000 or 10,000, and so on.*

For example, $0.3 = \frac{3}{10}$, and $0.37 = \frac{37}{100}$.

The only thing to figure out is *how many* zeros to use, which isn't very hard.

....

* *This doesn't include repeating decimals or irrational decimals, just* terminating *decimals—the ones that have a definite end. See page 155 for more on terminating decimals.*

Danica's Diary

CHOCOLATE MALT MADNESS

A few months ago, I bought a big bottle of organic chocolate syrup—I was thinking about how much fun it would be to make an old-fashioned chocolate malt! I went online and found a great recipe that dates back to the 1920s.

I saw that for one malt, I'd need $\frac{1}{8}$ cup of chocolate syrup. I was considering making enough chocolate malts for a group of my friends, so I wanted to know how many *total malts* I could make with my bottle of syrup. Since I knew that $\frac{1}{8}$ cup of chocolate syrup would make 1 malt, I also knew that 1 cup of chocolate syrup would make 8 malts. To figure out how many total malts I could make, I then just needed to figure out how many *total cups* of syrup were inside the bottle.

The only problem was, the bottle said it contained 1.03507 *liters* of syrup; I needed to know how many *cups* that equaled. Luckily, there are plenty of websites that can help with conversions* so I was able to find out that 1.03507 liters equals approximately 4.375 cups.

I knew how to measure out 4 cups but wasn't sure what 0.375 cups amounted to. I would have a better idea of how much syrup to use if it were expressed as a fraction, since that's what measuring cups use. So I converted the decimal number 0.375 into a fraction, following the easy steps below!

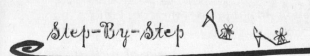

Step-By-Step

Converting decimals → fractions:

Step 1. *Count* the number of digits after the decimal point.

Step 2. Put the *same number* of zeros after a 1 (for example: 10; 100; 1,000; and so on): that's your denominator.

....
* *Just Google "measure" and "conversion" to find some!*

Step 3. *Drop* the decimal point from the original number: that's your numerator.

Step 4. Don't forget to reduce, if possible, and you're done!

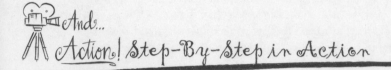

And...

Action! Step-By-Step in Action

Let's write 0.45 as a fraction.

Step 1. Count the **two** numbers after the decimal point: 0.**45**

Step 2. Make a denominator with **two** zeros after a 1: $\frac{?}{100}$

Step 3. Drop the decimal from the original number and make that the numerator: $\frac{45}{100}$

Step 4. Reduce: $\frac{45}{100} = \frac{45 \div 5}{100 \div 5} = \frac{9}{20}$

And our answer is $0.45 = \frac{9}{20}$. Done!

Take Two! Another Example

How about the chocolate syrup example from above? Let's turn 0.375 into a fraction.

Step 1. Count the **three** numbers after the decimal place: 0.**375**

Step 2. Make a denominator with **three** zeros: $\frac{?}{1000}$

Step 3. Drop the decimal from the original number and make that the numerator: $\frac{375}{1000}$

Step 4. Can we reduce? Looks like 25 divides into the top and bottom: $\frac{375}{1000} = \frac{375 \div 25}{1000 \div 25} = \frac{15}{40}$. Is it reduced yet? Nope—looks like another 5 can divide into the top and bottom: $\frac{15}{40} = \frac{15 \div 5}{40 \div 5} = \frac{3}{8}$

So, 0.375 cups = $\frac{3}{8}$ cups. And the bottle of syrup had 4.375 cups in it, so that means the bottle has $4\frac{3}{8}$ total cups of syrup in it. For each cup of syrup, I knew I could make 8 malts, so the 4 cups gave me $4 \times 8 = 32$ malts. Plus, I could get 3 malts out of the $\frac{3}{8}$ cups of syrup. That gave me a total of $32 + 3 = 35$ malts. Yum!

What's the Deal?

But, you say, what if the decimals are repeating? How can you count the number of places after the decimal when there is an infinite number of them?

Good question. For now, I'll show you a trick to convert repeating decimals into fractions when *all* the numbers after the decimal point are repeating, like $0.\overline{123}$.

Converting Repeating Decimals ➜ Fractions

"No no no no no no no no no no no no!" Have you ever had to babysit for a little brother or sister who screamed like this at the top of his or her lungs, over and over and over again?

How annoying! Actually, little kids *all over the world* do this, and when they do, many sound the same. Did you know that the word for *no* is very similar in tons of languages? Way more than any other word. Let's see, there's *No, non, nao, nyet, nej, nee, ne, nie, nem,* and *nu,* just to name a few. In Germany, you'd hear a little kid screaming, "*Nine nine nine nine nine nine nine nine nine nine!*"*

Can't you just picture some little kid screaming "9!" over and over again? That would be pretty funny. You could take away his ice cream and say, "How many pieces of broccoli would you like to eat?" And he'd scream, "9! 9! 9!" over and over again.

Of course, he wouldn't understand English, but it would still be pretty amusing. And you'd probably want to give him his ice cream back after a few seconds. He might throw something at you.

As you might have guessed, this little kid *repeating* "nine" has something to do with *repeating decimals.* When we convert repeating decimals to *fractions,* instead of putting a 1 followed by zeros in the denominator, can you guess what number we use instead? You got it: 9. The only trick is, you have to separate out the whole part of the number (if there is one) from the repeating decimal part. Here's how it works:

---

* *The word for "no" in German is actually spelled* nein *but it sounds just like "nine!"*

Step-By-Step

The 9s trick for converting <u>repeating decimals</u> to fractions:

Step 1. Split the number into its whole part (if any) and its decimal part. **Count** the number of repeating decimals (only if *all* the digits after the decimal are part of the repeating pattern).

Step 2. Put the **same number** of 9s in the denominator of the fraction you're converting it to.

Step 3. Drop the repeating bar from the original number's decimal part, and make that the numerator.

Step 4. Reduce. If there was a whole part in the original number, now's the time to combine it with our fraction—this will create a mixed number. Done!*

And... Action! Step-By-Step in Action

Let's convert $5.\overline{123}$ to a fraction.

Step 1. There is a whole part, so separate 5 from the decimal part. We'll just put 5 aside for now, remembering that $5.\overline{123} = 5 + 0.\overline{123}$. Continuing, count the **three** repeating decimals: $0.\overline{123}$.

Step 2. Put **three** 9s in the denominator: $\frac{?}{999}$.

Step 3. Drop the decimal and bar from the decimal part of the original number and make that the numerator: $\frac{123}{999}$.

Step 4. Reduce.

By adding up the digits of the numerator $1 + 2 + 3 = 6$, we

. . . .

* *If not all the digits after the decimal point repeat, then this trick doesn't work. You can still use it, but you have to be sneaky. For example, to convert $1.1\overline{4}$ to a fraction, you'd first multiply the decimal by 10 to get $11.\overline{4}$, then you could use the "9s trick" and when you got your mixed number, $11\frac{4}{9}$, you'd have to convert it into an improper fraction, $\frac{103}{9}$, so that you could divide the entire answer by 10—in other words, you could multiply just the denominator by 10, so you'd get $\frac{103}{90}$ as your final answer.*

can see that the top is divisible by 3, and so is 999 (remember the divisibility tricks on p. 9?). So let's divide by 3 on top and bottom and see where we stand: $\frac{123 \div 3}{999 \div 3} = \frac{41}{333}$. This is reduced, so we can combine it with the whole part: $5.\overline{123} = 5 + \frac{41}{333} = 5\frac{41}{333}$. Yikes! That's a pretty hairy-looking mixed number, but that's our answer: $5.\overline{123} = 5\frac{41}{333}$.

(How did we know that $\frac{41}{333}$ was reduced? This is tricky. Let's take apart 333. When we divide 333 by 9, we get 37, which, after a bit of factor work, turns out to be prime. So the prime factors of 333 are 3, 3, and 37. None of these divide into 41, so our fraction is indeed reduced!)

What's It Called?

Terminating Decimal

A **terminating decimal** is a decimal that ends; it doesn't repeat. So, 3.51 is a terminating decimal, and $3.5\overline{1}$ is *not* a terminating decimal, since the 1 repeats . . . forever. (To *terminate* actually means to "end." There's a famous trilogy of scary movies called *The Terminator* about evil robots that kill people, *ending* their lives. Kind of creepy, I know, but thinking about these movies always helps me to remember what terminating decimals are!)

"*I* have a lot of respect for genuinely intelligent girls who don't dumb themselves down." Kayleigh, 16

"*S*mart girls are cool. They are generally fun to be around and good to have conversations with." Mackenzie, 19

"*I* don't think dumb girls are dumb because they can't learn anything, but because they choose not to learn. If you are doing poorly in a class, you are not dumb—you just need extra help or need to try harder." Anna, 17

What Do You Have to Say?

Ahead of the Crowd

Here are some fraction/decimal relationships that come up a lot in real life (and in homework!), so it's nice to have them handy in a list. If you can *memorize* them, you'll be ahead of the crowd:

$$\frac{1}{2} = 0.5 \quad \frac{1}{4} = 0.25 \quad \frac{3}{4} = 0.75 \quad \frac{1}{3} = 0.\overline{3} \quad \frac{2}{3} = 0.\overline{6}$$

$$\frac{1}{5} = 0.2 \quad \frac{1}{10} = 0.1 \quad \frac{1}{100} = 0.01$$

Doing the Math

Change these decimals into fractions. Remember: If it terminates (ends), use zeros in the denominator, and if it repeats, use 9s. I'll do the first one for you.

1. 3.4 =

Working out the solution: Since there is just one number after the decimal, we'll put just one zero in the denominator: $\frac{34}{10}$. Now reduce: $\frac{34}{10} = \frac{34 \div 2}{10 \div 2} = \frac{17}{5}$. Done!

Answer: $3.4 = \frac{17}{5}$

2. 0.8 =

3. $0.\overline{8}$ =

4. 1.5 =

5. $1.\overline{5}$ = (Hint: don't forget to separate the whole part before continuing!)

(Answers on p. 291)

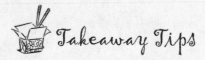

Takeaway Tips

- If the decimal ends, or *terminates*, count the number of digits after the decimal point and use that same number of zeros in the denominator of the fraction you're converting it to. The numerator is just your original number—without the decimal point!

- If the decimal *repeats*, and if all the digits after the decimal point are part of the repeating pattern, then count the number of decimal places under the bar, and use the same number of 9s in the denominator. Your numerator will be the original number, without the decimal point and without the bar. Don't forget to separate the whole part of the number (if there is one) before using the trick, and then combine it later as a mixed number!

- Always *reduce* your fractions at the end.

What's Your Math Horoscope?

We all approach math—and life—differently. To learn how a girl's sign can affect the way she tackles homework, tests, and more, I consulted expert astrologer Lilli Mulonas. Find your sign and see how you compare!

Aries: March 20 through April 19
Ruled by the "in-your-face" planet Mars, you have the energy, drive, and ambition to accomplish whatever you set your mind to. You want to win. No—you *have* to win! Your natural fearlessness and determination will serve you well in mathematics. Your biggest challenge in math will be a little thing called patience. You like things to happen *right now*, and oftentimes math requires time and, yes, patience, to master certain topics. When math gets frustrating, instead of getting fed up with it, tap into your competitive side. Are you really going to let a few little numbers get the better of you? No way! Say to yourself, "I can beat this—I know I can!" Also, homework doesn't have to be a drag: if the thought of sitting by yourself and working is getting you down, invite a couple of friends over for a group study session.

Taurus: April 20 through May 20
Taurus, you know how to slow down and smell the roses, and you love being surrounded by beauty and art. You may aspire to be a singer, architect, interior designer, or other profession geared toward making the world a more beautiful place! With such a focus on art, you may wonder, "Where does math fit into my life?" Well, architects and interior designers know the value of mastering math—they use it every day! Even music is extremely mathematical in nature. Look for the inherent sense of beauty in numbers and patterns—you'll love studying numbers that show up in nature, like pi or the Fibonacci sequence.

You also have a lot of patience, Taurus, which means that you have the ability to sit down and study when you need to. What's your biggest challenge in math? When Taureans feel rushed or stressed, they tend to have difficulty performing. So in a math test situation, it's very important for you, especially, to calmly look over the test, find which problems look easiest, and do those first. Then, approach the more difficult-looking ones. After you've warmed up on the easier ones, you may find the others aren't as hard as you feared!

Gemini: May 21 through June 20
Your life is a beehive of activity! The phone's ringing, the TV is on, music is blasting from somewhere in the back of the house, and your friends are chatting up a storm in the kitchen. You're great at avoiding your biggest turnoff: boredom. With the whole world making jokes about how "dull" math is, it's no wonder that there have been times when you've avoided it like the plague. You love solving a good mystery, though, Gemini, and if you can see math for what it is—a series of puzzles to solve—then you've got it beat, girl!

You are a quick study when it comes to math—but you don't do well in a room by yourself (you'd much rather be where the action is), so you often don't give yourself enough time to master a concept before moving on to other activities. Find some classmates to join you for group homework sessions. Read the problems and discuss them out loud. Talking things out always helps you in other areas of your life—why should math be any different? Language tends to be your forte, so anytime you're faced with word problems, you should feel confident in your ability to "translate" them into math language (see chapter 15). You are a natural born communicator—if you remember that, you can use it to your advantage for tons of success in math (and life)!

Cancer: June 21 through July 23
Ruled by the reflective moon, you're blessed with an excellent memory! So as long as you apply yourself, memorizing math rules and "step-by-step" problem-solving techniques should come fairly easy to you. Your biggest challenge in math is overcoming tension and worry, so when it comes to tests, do what you can to stay relaxed. You know and I know that you can pull this off, Cancer. When you're about to take a test, think positive thoughts, even if they seem silly—thoughts like, "I can't wait to take this test! Yes, bring it on!" The sillier the better. And do what you can to make your homework time "cozy." Wear bunny slippers and make yourself some warm herbal tea to sip while you study. Put a soft pillow on your chair. Create an

environment that feels welcoming and homey. By the way, you'll need your math skills to count your money someday—Cancer outnumbers the rest among the very richest of us all!

Leo: July 21 through August 21

Lights! Camera! Action! Your friends would describe you as bubbly, outgoing, and larger than life. You're destined for greatness, and everyone knows it. What most people may not realize is that you have a true gift for organization. You are great at focusing on your work—if you believe it is worth your time. Math homework might not be your favorite activity, but think of the big picture: As you sit down to study, visualize yourself graduating from high school with honors, resplendent in cap and gown, with friends, family, and all who admire you sitting in the front row, applauding your success—marveling at the intelligent, savvy young woman you've become and snapping pictures as you receive your diploma. And think of the designer suits you'll wear someday as you click-clack in your heels to your high-powered job. Beauty and brains—you can have it all! Let this image inspire you and propel you toward an A+. So, set the stage: Background music—*check*. Books neatly stacked on your bedroom desk—*check*. Ready, set, go. You're on!

Virgo: August 23 through September 22

You're conscientious and organized, Virgo. You love having a plan, and you love checking off items from your "to do" list. Math probably comes more naturally to you than to most of your friends, because you were practically born with an understanding of numbers and order and data—plus, you're very observant, so you'll catch details that most people miss. (A wonderful quality to have!) Being prepared is important to you—you always make time for your studies, because you know that if you don't you'll have a tough time sleeping that night, worrying about unfinished work. Virgos' biggest challenge is not getting hung up on small, nitpicky details. When doing your math homework or taking a test, try to think about the big picture—don't get stuck on which eraser to use or if the page gets smudged or making your handwriting perfect. Neat is good, but perfect is unnecessary and time-consuming. Also, remember that while getting an A is great, a B+ really is a good grade, too. Don't sweat the small stuff, Virgo—you have a lot to be proud of!

Libra: September 23 through October 23

Libra's symbol is the scales: you have a need for and understanding of balance, in all areas of your life, and you are easily able to see both sides of every situation. Ruled by Venus, you also have a deep appreciation for beauty. Libras make great judges, lawyers, art dealers, boutique owners, architects, and more—and math plays into many of these careers! With such an innate understanding of balance, you will likely have an appreciation for the idea of algebra— even if you don't get the details of it right away. After all, "solving for x" revolves around taking an equation—in which both sides are equal—and performing operations on both sides, while keeping the scales balanced, until x's value is revealed (see p. 241 for more on algebra). In terms of studying? While you'd rather spend your time picking out just the right dress for your friend's party tonight than doing math homework, your sense of balance tells you that studying is important. So, head straight home after school and get down to work, take a quick shower, slip into your party dress, and off you go for a well-deserved fun time with your friends.

Scorpio: October 24 through November 22

If there's a puzzle or riddle, Scorpio, you won't rest until you've solved it; you're a born investigator! You love to explore, search, and dig for answers. Someday, you may join the scientists studying DNA, now so widely used in solving crimes, or become a financial analyst or stockbroker. You can bet they all use math every day. You've got a clever mind, and a private one. Your classmates may not know much about your innermost thoughts. But this is not because you're hiding anything—you're just a private sort of girl. Because of this, you shouldn't have much of a problem with homework—Scorpios are independent gals, so who needs a partner, anyway? (You do love the idea of having a romantic partner, but my-my, you're very choosy— good for you!) However, you can be stubborn at times, and one of your biggest challenges in life is learning to trust others—so when you have trouble understanding a math concept, don't stay stuck. Go ahead and ask for help! You might be surprised by what your teacher—or others—have to offer.

Sagittarius: November 23 through December 21

Jovial, joyous, generous, happy-go-lucky, optimistic Sagittarius! People love having you around, and you love to have fun. You're also very smart! The only problem is, you may not want to *stop* having fun and exercise that brain of yours. As with Geminis, boredom is a total turnoff for you, so, in order to get things (like homework) done, you'll need to create a kind of fun versus responsibility formula; you'll have to make deliberate plans to work before you start chatting with your friends again. However, whenever you have studying or homework to do, be sure to make fun plans for afterward as a reward for getting your work done! Overdoing one or the other won't work for you—you'll get burned out, or start to feel guilty about what's not getting done. Here's a tip: Find a sport, outside school. Tennis, golf, softball, volleyball, hiking—anything active. These activities are essential for you freedom-loving gals and will make it easier for you to sit down and study when you need to. Another defining Sagittarius trait is that you're attracted to anything that feels like a game, and math is a series of puzzles to solve, after all. If you think of math in this way, it may make it easier for you to focus on it when you need to—and maybe even enjoy it!

Capricorn: December 22 through January 20

There is no sign more capable of discipline and self-control than Caps! Surely, you were born mature, with an innate sense of responsibility, and you take your studies very seriously. You don't need a study group for motivation—you're one of the few signs that does very well by herself. Your biggest challenge is not worrying when you don't understand things. Worrying never helps anything. When you start to worry, try singing a cute little song inside your head like "Don't Worry Be Happy," and applaud yourself for what you *have* accomplished. Just be as patient as you can while learning new concepts (Capricorns have a lot of patience), and know that you're not supposed to understand everything perfectly the first time. With your diligence, you'll find the help you need in this book and from your teachers. (That said, sometimes Caps care about what others think too much—don't be shy about asking for help!) Your ambition is powerful, and it'll earn you much success in everything you set your mind to, including math. And when you start scoring high on tests, you won't start bragging about your As. It's with that spirit that you'll enter the adult world someday and reach the very top of the mountain of your choice!

Aquarius: January 20 through February 18

Your ruler is Uranus, the planet which demands independence and freedom of choice. That is a tall order! (Also, be careful how you choose your friends. Friends are often a reflection of ourselves—and this is especially true for your sign—so hang around people you *admire*, not just those you have fun with.) Your love of freedom will often tempt you to ignore your homework and go to the mall instead, but remember the big picture: As an adult, freedom costs money! As you grow up, getting good grades and really learning what you are taught in school—especially in math—is the key to the ultimate independence you'll seek all your life. Do you know how many students end up choosing majors in college based on the ones that require the least math? I had a friend in college who always wanted to be a doctor, but he didn't want to have to take calculus—and now he'll never be able to pursue that dream. Math really is power, and freedom. The sky's the limit for you: you are bright enough to master math and have all the freedom to be and do what you want in the world later on.

Pisces: February 19 through March 21

You have a love of music and magic—and the ocean! After all, yours is not only a water sign, but has the symbol of two fish! Whenever possible, take your homework to the sea or lake or river—or you can just listen to the sounds of the ocean on your iPod. Putting yourself in touch with water will awaken your mind and inspire your imagination. And there really is magic in numbers—just Google "math magic," and you'll find tons of tricks you can amaze your friends with! Also, use your vivid imagination to think up creative ways to remember math concepts. You'll likely be attracted to mnemonic strategies like the "MAD Face" method of converting mixed numbers into improper fractions (see p. 45) or the Sparky the dog method for remembering which part of a division problem is the divisor, quotient, and dividend (see p. 126) or any that you *create* for yourself. You have the ability to think outside the box, so go for it! A little-known Pisces named Albert Einstein once said, "Imagination is more important than knowledge." He certainly used his imagination to look at math in an entirely new way—and so can you!

Chapter 13

Sale of the Century!

Converting Percents to and from Decimals and Fractions

*D*ecimals and fractions aren't the only types of numbers you'll need to master in order to take full control of your shopping experience (and, yes, your math homework . . .). You'll also need to understand **percents.**

Percents are everywhere!

In order to answer a question like "What is 75% of $60?" you first need to *convert* the percent into a fraction or a decimal, and then multiply that fraction or decimal times the dollar amount: we *never* multiply or divide percents while they are still in "percent" form.

Let's think about the word *percent* for a second. Technically, *percent* means "parts per hundred." If you have 19%, you have 19 "per" cent. And *cent* means "100," which is why a century is 100 years. This is why 19% = 19 per 100 = $\frac{19}{100}$.

We'll come back to this in a few pages. For now, suffice it to say, if we want to find out how much that halter top on the preceding page costs, we need to know how to go back and forth between percents and decimals, and also between percents and fractions. Let's do it!

Converting Percents to Decimals (and Vice Versa)

Going back and forth between percents and decimals is mostly very easy. All you do is add or take away the % sign, then move the decimal point two places—that's it! Here are some examples:

$$45\% = 0.45 \quad 83\% = 0.83 \quad 99\% = 0.99$$
$$3\% = 0.03 \quad 245\% = 2.45 \quad 70\% = 0.7$$

(Note: 0.7 has the same value as 0.70, but we don't need that extra zero, which is why it was dropped.)

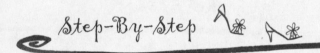

Step-By-Step

Converting percents to decimals:

Step 1. Drop the % sign.

Step 2. Move the decimal point to the *left* two places. Done!

And...
Action! Step-By-Step in Action

Convert 60% to a decimal.

Step 1. Drop the % sign: 60

Step 2. Move the decimal point to the *left* two places. Since 60 = 60.0, we know where the decimal point is. Now, moving it to the *left* two places, we get 0.60, which is the same as 0.6, since we don't need that extra zero at the end.

Answer: 60% = 0.6

 Take Two! Another Example!

Convert 2% into a decimal.

Step 1. Drop the % sign: 2

Step 2. Move the decimal point to the *left* two places. Since 2 = 2.0, we know where the decimal point is, and moving it to the *left* two places, we get 0.02.

Answer: 2% = 0.02

 Take Three! Yet Another Example!

Convert 1.3% from a percent into a decimal.

You might be thinking, "But it's already a decimal!" Well, yes, sort of: it's a percent that happens to include a decimal. But we need to write the value as a decimal *without* a percent sign, and that means we need to follow the same step-by-step method we've been using.

Step 1. Drop the % sign: 1.3

Step 2. Move the decimal point to the left two places: 1.3 → 0.013

Answer: 1.3% = 0.013

See? It works out fine!

Now we'll go in reverse, and change decimals into percents . . .

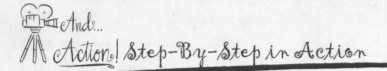

Step-By-Step

Converting decimals to percents:

Step 1. Move the decimal point to the *right* two places.

Step 2. Add a % sign. Done!

And... Action! Step-By-Step in Action

Convert 0.72 to a percent.

Step 1. Move the decimal to the *right* two places: 0.72 → 72.

Step 2. Add a % sign: 72%

Answer: 0.72 = 72%

Take Two! Another Example!

Convert 0.3 to a percent.

Step 1. Move the decimal place to the *right* two places: 0.3 = 0.30 → 30.

Step 2. Add a % sign: 30%

Answer: 0.3 = 30%

Watch Out! Always count your decimal points and zeros carefully!

Many people would be tempted to say 0.3 = 3%, for example, but they'd be *wrong*. If you count your decimal places carefully, you'll find that 0.3 = 30% (as we showed above), and 3% = 0.03; don't skip steps, and you'll avoid common mistakes like this.

Doing the Math

Convert the following percents to decimals and decimals to percents. I'll do the first one for you.

1. 100% =

Working out the solution: First, we drop the % sign, and then we move the decimal point two places to the left: $100 \to 1.00 = 1$. Yep, the "decimal" form of 100% is 1!

Answer: 100% = 1

2. 5% =

3. 0.75 =

4. 500% =

5. 0.09% =

6. 1.44 =

7. $\frac{1}{2}$% = (Hint: this is "half of a percent," not "one half." First rewrite $\frac{1}{2}$ as 0.5 and then continue.)

(Answers on p. 291)

QUICK NOTE! You may be thinking, "Yeah, going back and forth between decimals and percents is easy; I basically just move the decimal point two places. But how will I remember *which way* to move the decimal point?"

Think about the alphabet. If you wanted to go from the letter *D* (which stands for *decimals*) to the letter *P* (which stands for *percents*), you'd move to the *right*, because the *P* is to the right of the *D* in the alphabet—right? You also move the decimal point to the *right* when you convert from a decimal to a percent. And if you wanted to go from the letter *P* to the letter *D*, you'd move to the *left*. So you can picture the alphabet, and "see" where the *D* and the *P* are, if that helps you!

Also, you probably already know that 50% = 0.5. So, if you ever get confused about which direction you're supposed to move the decimal point when converting decimals to percents, just imagine that you are converting 0.5 to 50%. And if you are converting percents to decimals, imagine that you are converting 50% to 0.5—that should keep things straight for you!

Converting Percents to Fractions (and Vice Versa)

You're shopping (again), and browsing through a couple of clothing racks in one of your favorite stores. One rack says "20% Off!" and the other says "$\frac{1}{5}$ Off!" Which is offering the better deal?

There are lots of situations in life (and in homework) where you'll want to compare fractions to percents, so it's helpful to know how to convert one to the other, and vice versa!

Percents → Fractions

You may have been wondering (okay, you probably weren't wondering) why I've held off on presenting the formal definition of *percent*. Well, it's because I wanted to wait until we got to the *fractions* section of this percents chapter . . .

"ring ring"

What's It Called?

Percent

A *percent* is a fraction whose denominator—which is always 100—is expressed as the % sign. For example: $19\% = \frac{19}{100}$

Percents are just *fractions* with denominators of 100. That's it!

For example, $43\% = \frac{43}{100}$

Like I mentioned at the beginning of the chapter, the word *cent* means 100. That's why a century = 100 years.

$$19\% = 19 \text{ "per" cent} = 19 \text{ "per" } 100 = \frac{19}{100}$$

In fact, the little % sign sort of looks like a little fraction, with two little zeros for 100! It's like the % sign and the 100 are interchangeable—you can use one or the other to express the same amount.

So when you see a percent and you want to convert it into a fraction, all you have to do is drop the % sign, and put the number on top of a fraction with 100 on the bottom; it means the *same thing*. You're now just expressing it as a fraction, instead of as a percent.

Step-By-Step

Converting percents → fractions:

Step 1. Drop the % and make a fraction with the number on top and 100 on the bottom.

Step 2. Simplify/reduce the fraction, and you're done!

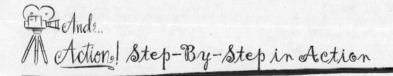 **And... Action! Step-By-Step in Action**

Let's convert 24% to a fraction.

Step 1. Drop the % and put the number "over" 100: $\frac{24}{100}$

Step 2. Reduce the fraction: $\frac{24}{100} = \frac{24 \div 4}{100 \div 4} = \frac{6}{25}$. Since the prime factors of 25 are 5 and 5—and 5 doesn't share any factors with 6—then we know the fraction is reduced.

Answer: $24\% = \frac{6}{25}$

Sometimes percents have decimals or fractions *in* them, and look like this: 7.5% or $8\frac{1}{2}$%, and you might be asked to convert these into fractions.* Then what? No worries. Even if they have decimals or fractions *in* them, you still just have to put the whole thing on top of a fraction with 100 on the bottom. (Remember, that's what the little % sign means!)

 Take Two! Another Example!

Let's convert $8\frac{1}{2}$% into a fraction, step-by-step.

Step 1. What do we do? We put $8\frac{1}{2}$ over 100 and drop the % sign: $\frac{8\frac{1}{2}}{100}$. Now, following the "complex fraction" methods (see p. 96), we should convert the mixed number into an improper fraction, and then write the denominator as a fraction, too, to make it "tall and skinny": $\frac{\frac{17}{2}}{\frac{100}{1}}$.

Use the Means and Extremes method to simplify this into:

$$\frac{\frac{17}{2}}{\frac{100}{1}} = \frac{17 \times 1}{100 \times 2} = \frac{17}{200}$$

Step 2. Can we reduce this? Nope—17 and 200 don't share any factors, because 17 is prime and does not divide evenly into 200.

Answer: $8\frac{1}{2}$% $= \frac{17}{200}$

Learning to be flexible in converting numbers from one form to another is a great math skill that'll come in handy all over the place. It's all about understanding that there are lots of ways of writing the same value.

---

* We'll do examples of fractions with decimals in them in chapter 14.

Doing the Math

Convert these percents to fractions. I'll do the first one for you.

1. $\frac{1}{2}\% =$

<u>Working out the solution</u>: Remember, this is "half of a percent," not "one half." We have to think of this as a percent. First, drop the % and put "the whole darn thing" over 100: $\frac{\frac{1}{2}}{100}$. Then rewrite the denominator as a fraction $\frac{\frac{1}{2}}{\frac{100}{1}}$, so we can do "means and extremes" (see p. 97) on this "tall and skinny" complex fraction.

$$\frac{\frac{1}{2}}{\frac{100}{1}} = \frac{1 \times 1}{100 \times 2} = \frac{1}{200}$$

<u>Answer</u>: $\frac{1}{2}\% = \frac{1}{200}$

2. $25\% =$

3. $\frac{1}{5}\% =$

4. Let's figure out the price of that halter top from the beginning of the chapter. What is 75% of $60? (Hint: First convert 75% to a fraction. And remember: *of* means "multiply" when immediately surrounded by two numbers.)

(Answers on p. 291)

Fractions → Percents

Now I'll show you how to do this in reverse, and change fractions into percents!

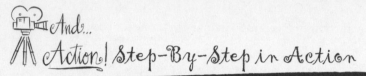

Converting fractions → percents:

Step 1. First convert the fraction to a decimal. (See the section on converting fractions → decimals on p. 137 if you want a refresher.)

Step 2. Now it's just a regular decimal → percent conversion, so move the decimal point two spaces to the right, and add a % sign. That's it!

And...

Action! Step-By-Step in Action

Let's convert $\frac{1}{4}$ into a percent.

Step 1. First, convert $\frac{1}{4}$ into a decimal. (See p. 125 for a review of decimal division.) We tip the fraction over and divide $4\overline{)1.00} = 0.25$

Step 2. Move the decimal point two places: 25 and add the % sign: 25%

You probably already knew that $\frac{1}{4}$ = 25%, but now you've proven it!

Take Two! Another Example!

Let's convert $2\frac{2}{5}$ into a percent.

Step 1. First, convert $2\frac{2}{5}$ into a decimal.
We know that, by and large, the easiest way to work with mixed numbers is to convert them into improper fractions, so we get: $2\frac{2}{5} = \frac{12}{5}$. And to convert this to a decimal we just tip it over and divide:

$5\overline{)12.0}$ with quotient 2.4. So, $2\frac{2}{5} = 2.4$.

Step 2. Move the decimal point two places to the right: 240, and add a percent sign: 240%.

Answer: $2\frac{2}{5} = 240\%$

QUICK NOTE! Note that, when converting fractions →
percents, if the original number is *bigger* than 1, then when
it's written as a percent it will be *bigger* than 100%.

Doing the Math

Convert these fractions to percents. I'll do the first one for you.

1. $\frac{1}{5} =$

Working out the solution: First let's make this a decimal
by tipping and dividing the fraction: $5\overline{)1.0}$ with 0.2 above. Then we
move the decimal point two places: 20, and add a %
sign: 20%.

Answer: $\frac{1}{5} = 20\%$

2. $\frac{1}{2} =$

3. $\frac{3}{2} =$

4. $4 =$ (Hint: First rewrite 4 as the fraction $\frac{4}{1}$, then proceed
normally.)

(Answers on p. 291)

QUICK NOTE! When converting from fractions → percents, a great way to check your work is to convert
your answer back to a fraction again and see if you
get the fraction that you started with.

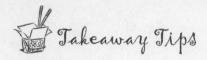
Takeaway Tips

- *Going back and forth* between percents and decimals is simply a matter of moving the decimal point two places, and either adding or dropping the % sign.

  Decimals → percents: we move the decimal point two places to the *right*. Percents → decimals: we move the decimal point two places to the *left*.

  (If you forget which direction to move the decimals, just remember that 50% = 0.5, or think about the alphabet—how you move to the *right* to go from D to P and to the *left* to go from P to D!)

- Converting *percents → fractions* is simple: you put the "percent number" over a denominator of 100, then drop the percent sign. And don't forget to reduce!

- To convert *fractions → percents*, we first convert fractions → decimals, then decimals → percents.

Danica's Diary

MATH QUIZ FREEZE-UP (AND A GUARDIAN ANGEL)

The seventh grade: a new school for me.
Everything seemed to be going okay; I'd made a few friends already and hadn't embarrassed myself too horribly when I showed up for English class still wearing my gym clothes. I was beginning to think I might survive this new school. But I was in for an unpleasant surprise: suddenly, math got much, much harder.

The math at my new school just seemed different than the math I'd learned in sixth grade, and the new challenges scared me. Adults had always referred to math as "hard," and now I understood why! Our teacher would write things on the board that just didn't make sense. I felt completely lost.

I never would have guessed that my failure to keep up could have been in any way the fault of the teacher. At that point, I felt defeated and had decided that I couldn't do math.

Midway through the year, the school switched our math teacher to a gentle, more experienced woman named Mrs. Jacobson. On one of the first quizzes she gave us, I can remember staring at it, seeing nothing but a blank page. You might remember this story from the introduction to this book. It was truly one of the most terrifying moments of my life! I had a horrible, sinking feeling in the pit of my stomach. I just sat there as the minutes ticked on, having *no idea* how to approach these problems. I had to hold back tears as I heard the bell ring for recess, knowing that I had failed again. It was awful.

But here's where the story gets interesting: For reasons I still do not truly understand, Mrs. Jacobson did not collect my quiz at the end of class. Instead, she let me stay through recess after all the other kids had left. I remember thinking, *Is this fair? Why do I get extra time?* But when I looked at her, she just smiled at me. It's like she was saying, "I know you can do this." Somehow, I relaxed, and was able to do some of the problems. I actually scored a C+ on that quiz, and I quickly became an A student in that class. Funny how just *relaxing* can make such a difference.

I might never know why Mrs. Jacobson chose to reach out to me just then. All I know is that her faith in me at that moment is a gift I'll never forget.

Chapter 14

A Choreographed Performance

Mixing Fractions, Decimals, and Percents Together

$\mathcal{I}$t's helpful to be able to go *back and forth* easily between fractions, decimals, and percents, because all throughout school—high school and beyond—they'll keep coming back, over and over again.

Think about dance class or gymnastics, when you learn isolated moves, and then, at some point, you put them all together into a choreographed routine. It's one thing to do a series of turns to the corner of the room or to practice your high kicks. It's another to do one of those turns and do a high kick coming out of it!

So how do we combine some of the fraction, decimal, and percent "moves" we've learned thus far into a beautifully choreographed routine? Well, we've already combined some of these moves in previous chapters. You've seen things like:

$$\frac{\frac{1}{6}}{\frac{3}{4}} \text{ and } \frac{\frac{1}{4} + \frac{1}{2}}{2 - \frac{1}{8}} \text{ and } \frac{1}{2}\% \text{ (one half of a percent)}$$

But what about $\frac{0.45}{5}$ or $\frac{0.6}{2.4}$?

When Fractions Have Decimals in Them

Let's focus on these types of fractions, because they'll show up in word problems and algebra, often when you least expect it. How do we handle them? No problem!

Remember, fractions are division problems in disguise, so you can always just tip them over and divide them, like this:

$$\frac{0.45}{5} = 5\overline{)0.45}$$

But decimal division can be tedious—and sometimes you'll want to end up with a fraction answer, not a decimal. So what's a gal to do? Personally, I prefer to handle these kinds of fractions by making them more friendly (i.e., by getting rid of the decimal point!)

For example, with $\frac{0.45}{5}$, there are **two** places to the right of the decimal point, and so we multiply the top and bottom by **100**, because the **two** zeros will get rid of the **two** decimal places.

$$\frac{0.45}{5} \times \frac{\mathbf{100}}{\mathbf{100}} = \frac{0.45 \times \mathbf{100}}{5 \times \mathbf{100}} = \frac{45}{500}$$

Now that the decimals are gone, we can reduce it, using the GCF of 45 and 500, 5.

$$\frac{45}{500} = \frac{45 \div \mathbf{5}}{500 \div \mathbf{5}} = \frac{9}{100}$$

Answer: $\frac{0.45}{5} = \frac{9}{100}$

So that's how I like to handle fractions that have decimals *in* them. And, *hmm.* This looks an awful lot like 9%, doesn't it? Or, written as a decimal, 0.09. It's amazing how many different ways there are to express the *exact same* value.

Step-By-Step

Dealing with decimals in fractions:

Step 1. Look on the top and bottom of the fraction, and count the *biggest* number of decimal places (one, two, three, etc.) to the right of the decimal point.

Step 2. Multiply the top and bottom by a 1 followed by *that many* zeros—**10**, **100**, **1,000**, etc.—so that you get rid of the decimal. Remember to multiply top and bottom by the *same* number!

Step 3. Reduce as usual. If you want a decimal answer, of course, you can convert the fraction by tipping it over and dividing!

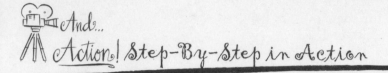
And… *Action!* Step-By-Step in Action

Simplify this expression into a nice-looking, reduced fraction.

$$\frac{0.35 + \frac{1}{2}}{5.5}$$

There are a few different ways to do this, but they all start with *first simplifying only the top part of this complex fraction.* Let's convert $\frac{1}{2}$ to a decimal so we can easily add it to 0.35. (We could also convert 0.35 to a fraction and then find a common denominator to add them together, but that would take longer.) As you probably know, $\frac{1}{2} = 0.5$. So, $0.35 + \frac{1}{2} = 0.35 + 0.5 = 0.85$. Now our fraction looks like: $\frac{0.85}{5.5}$. And now, we can apply our step-by-step method.

Step 1. On top, there are **two** places to the right of the decimal point; on the bottom, there is only one.

Step 2. So, if we multiply the top and bottom by **100** (**two** zeros will get rid of **two** decimal places), we'll take care of both pesky decimal points at the same time!

$$\frac{0.85}{5.5} = \frac{0.85 \times \mathbf{100}}{5.5 \times \mathbf{100}} = \frac{85}{550}$$

Step 3. Now that the decimal points are gone, we can reduce. We know that 5 is a factor of both 85 and 550, since both numbers end in a 5 or 0 (see p. 9 for divisibility tricks).

$$\frac{85}{550} = \frac{85 \div \mathbf{5}}{550 \div \mathbf{5}} = \frac{17}{110}$$

Answer: $\frac{0.85}{5.5} = \frac{17}{110}$

QUICK (REMINDER) NOTE! Remember our copycat fractions from earlier? Well, when we multiply the top and bottom of a fraction by 10, 100, 1000, etc., we're really multiplying the fraction by a copycat: $\frac{10}{10}, \frac{100}{100}, \frac{1000}{1000}$, etc. And since all copycat fractions equal 1, we aren't changing the *value* of the fraction—we're expressing the *same value.*

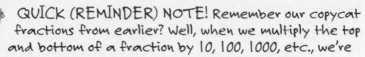 *Take Two! Another Example!*

Let's convert 7.5% into a fraction.

Just like always, to get rid of a percent, drop the % sign and put the entire thing over 100: $\frac{7.5}{100}$.

And now we can apply our step-by-step method: Let's multiply by a copycat fraction that will make that decimal go away. We only need to move the decimal point **one** place, so let's use **10**.

$$\frac{7.5}{100} = \frac{7.5 \times \textbf{10}}{100 \times \textbf{10}} = \frac{75}{1000}$$

Okay, at least the decimal's gone. Now we need to reduce it. Let's see:

$$\frac{75 \div \textbf{25}}{1000 \div \textbf{25}} = \frac{3}{40}$$

Hmm. 3 and 40 don't share any common factors, so we're done!

Answer: $7.5\% = \frac{3}{40}$

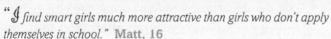

What Do You Have to Say?

Doing the Math

Make these expressions look more "friendly"—turn each into a decimal, a percent, and also into a "nice," reduced fraction without decimals in it. I'll do the first one for you.

1. $\frac{0.6}{5.4}$

<u>Working out the solution</u>: To make this a decimal, let's divide $5.4\overline{)0.6} = 54\overline{)6.000} = 0.1111111\ldots = 0.\overline{1}$ (See p. 143 for a review of repeating decimals.) To make this a percent, we just move the decimal point two places to the right and add a percent sign: **11.$\overline{1}$%**.

Next, let's convert $\frac{0.6}{5.4}$ into a "nice" fraction (without decimals). Since on top and bottom, there's only **one** digit to the right of the decimal point, we'll only need to multiply top and bottom by **10** to get rid of the decimal: $\frac{0.6}{5.4} \times \frac{10}{10} = \frac{0.6 \times 10}{5.4 \times 10} = \frac{6}{54}$. That looks much better. Can we reduce? Yep! $\frac{6 \div 6}{54 \div 6} = \frac{1}{9}$.

<u>Answer</u>: $\frac{0.6}{5.4} = 0.\overline{1}$, $11.\overline{1}$%, and $\frac{1}{9}$

2. $\frac{0.72}{9} =$

3. $\frac{0.1}{0.02} =$

4. $\dfrac{\frac{1}{2} + 0.4}{1\frac{1}{2} - 0.3} =$

5. $2.5\% =$

6. $0.4\% =$

(Answers on pp. 291–2)

Comparing Decimals, Fractions, and Percents

Can you put these numbers in the correct order, from least to greatest?

$$1.24, \ 99\%, \ \frac{5}{4}$$

The fastest way to compare different types of numbers like this is *first to convert them all to decimals*, and then compare them. Okay, 1.24 is already in decimal form. For 99%, just move the decimal point two places to the left and drop the % sign: 0.99. And $\frac{5}{4}$ = "5 divided by 4" = $4\overline{)5.00}$ with 1.25 above. (See p. 137 for a review of fractions → decimals.)

And now it's easy to put them in order from least to greatest: 0.99, 1.24, 1.25. So the answer is, from least to greatest: 99%, 1.24, $\frac{5}{4}$. Voilà!

Step-By-Step

Comparing fractions, decimals, and percents:

Step 1. Convert all numbers to decimals.

Step 2. Compare the decimals, using the "hand covering" technique on page 117. Done!

Doing the Math

Put these numbers in order from least to greatest. I'll do the first one for you.

1. $0.385, \frac{3}{5}, 39.5\%$

<u>Working out the solution</u>: Okay, 0.385 is already in decimal form. And $\frac{3}{5}$ = "3 divided by 5" = $5\overline{)3.0}$ with 0.6 above. So $\frac{3}{5} = 0.6$ and $39.5\% = 0.395$

Now we need to compare 0.385, 0.6, and 0.395. Using our decimal comparison "hand covering" technique, we get that the order from least to greatest is: 0.385, 0.395, 0.6

<u>Answer:</u> $0.385, 39.5\%, \frac{3}{5}$

2. $\frac{1}{6}$, 0.19, 16%

3. $\frac{7}{4}$, 200%, $1\frac{4}{5}$

4. 0.889, 89%, $\frac{8}{9}$

(Answers on p. 292)

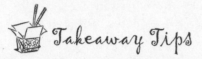 Takeaway Tips

- If you see decimals in your fraction, and you want to get *a decimal answer*, just divide the top by the bottom.

- If you see decimals in your fraction, and you want a *fraction* answer, just multiply top *and* bottom by 10, 100, or whatever you need to in order to get rid of the decimal points. (And then, of course, reduce.)

- To *compare* fractions, decimals, and percents, change them all to decimal form, then compare the decimals. I've found this to be the easiest, fastest way to do it!

 Danica's Diary

DUMBING OURSELVES DOWN FOR OTHER PEOPLE

You might think that dumbing yourself down for *anyone* sounds ridiculous. Believe it or not, though, people do it all the time, and not just in math class. It's too bad, but people often make themselves seem "less than" their real selves, in all sorts of ways, in order to avoid embarrassment—and unfortunately, I'm no exception.

I was 13 years old, and had been filming *The Wonder Years* for a few months already. One afternoon, I got the script for the newest episode, and it turned out to be an episode all about making fun of people behind their backs. To my horror, this included making fun of me!

The main character had to pick on me throughout the whole episode. He called me "Goody Two-shoes," and one of the specific things he made fun of was my good posture. The script had one character say that I walked around like I "still had the hanger" in my shirt.

Reading this, I got a horrible tingly sensation in my face and the back of my neck, like I wanted to pass out. I was *so* embarrassed, and tried to laugh it off when we were shooting the scene. But who was I fooling? I was mortified.

I know what you're thinking: "But they weren't making fun of you, they were making fun of your *character*." Yet, I ask you now—how could it have been anyone other than *me* they were making fun of?

I'd noticed in the past that I did sit up straighter than many of my friends. And even though good posture is supposed to be a positive thing, suddenly I was being made fun of for it, and all I wanted was to be "less than" perfect. I just wanted to fit in and be like everyone else!

In junior high, it's hard not to compare ourselves to our friends—we've all done it. You always hear, "It's important to be yourself and not care what others think; to not go along with the crowd." This is true, and it will help you to avoid potentially harmful peer pressure. But sometimes, with the little things, it's easy to forget our individuality, and we're tempted to give in to what "everyone else" is doing because it feels, well, *comfortable*.

My friends and I used to sit under a big tree at school and eat lunch. Lots of my friends hunched over or slouched while they ate. After reading the television script, I began thinking that they looked cooler than me, and that I seemed like a Goody Two-shoes in comparison. (Years later, I asked one of my friends about this, and she said, "What? No one thought that!" But at the time, I was *sure* everyone noticed whatever was "different" about me, and judged me for it.)

I actually began *working* at slouching over when I sat with my friends during lunch, while sitting in class—while sitting anywhere! Wherever I could practice slouching, I did. That way, by the time the "making fun of Danica" episode aired on television, people would just shrug and say, "But she doesn't sit up straight in real life; she slouches just like the rest of us." I continued creating the bad habit of slouching for months; I was convinced that I was more likeable that way.

Of course, this was *all* in my mind. Looking back, I'm sure no one even noticed my new, bad posture. And after a few years, I started noticing that my back would ache if I sat in one slouched position for too long. I certainly wasn't going to get any "cool" points for having a back that hurt.

At a certain point in school, it becomes "cool" to be less than what we can be—an underachiever—in whatever category is important to us at the time: bad posture, bad grades, bad attitude. But this stuff only seems cool for a very short period of time, and steps in the wrong direction are hard to reverse later on—believe me.

Of course, *flaunting* what you are good at isn't a great idea, either. When I was in the sixth grade, I remember getting a history test back that was marked A+. I was so excited, I pumped my arm up and down and said to myself, "Yes!"

Unfortunately, I said it loud enough for some of my classmates to hear. It was an innocent mistake, but I probably made some of them feel bad . . . and it's probably why my boyfriend at the time stopped speaking to me, come to think of it.

You could say, "Well, that's their problem for not studying more." But it's important to be sensitive to other people's feelings. Nobody likes a show-off, except maybe your parents!

You *can* be your best, without making others feel bad. In fact, by being humble and not bragging when you accomplish something you are proud of, you'll probably end up being an inspiration to the people around you, without even knowing it.

Chapter 15

The Universal Language
of Love . . . and Math

Introduction to Word Problems
and "Percent Of"/"Percent Off"

Translating from English to Math

*A*h, the dreaded **word problem**. Let me tell you something: succeeding at word problems is really about one thing—learning how to translate English into math.

You already know how to translate English in other ways. For example, if a cute guy text messages you, "Hey, whatcha doin?" what are the chances that he's *really* just interested in what you are doing at that moment?

Slim, right? The reason you get excited and maybe even blush is because you're *translating* that message into what it really means: "Hey, I kinda like you, and I want to know what you are doing *because* I like you." It's all in the translation.

In the language of math, we have an extended alphabet that includes numbers and symbols, and we put them together to form "statements" or "equations," like $1 + 3 = 4$. This is a much faster way of saying, "When you add one plus three, you get four. The language of math offers a more efficient way to write down certain ideas.

So what does this have to do with word problems? Everything. In word problems, we translate *from English into math*, solve whatever math problem is involved, then take the answer and translate it *back into English*. It's the process of *translating* that is usually the hardest part of the problem.

For example, a word problem might state: "Taylor has 12 magazines. Madison has two-thirds as many magazines as Taylor. How many magazines does Madison have?"

This word problem is asking us to find out what *two-thirds of 12 is*, right? The first step is to *translate* the English into math, like this: Two-thirds of 12 magazines = $\frac{2}{3}$ "of" 12, and that equals $\frac{2}{3} \times 12$. Now that we've translated it into "math," we can solve it: $\frac{2}{3} \times 12 = \frac{24}{3} = 8$. And next we translate back into English: Madison has 8 magazines.

The Universal Language

𝔇id you know that you can go to pretty much any country in Europe and North and South America, and many in Africa and Asia, and the people there will be using the same numbers and math symbols that you do? There are many towns and villages across the globe where no one speaks a word of English—yet, if you were dropped into one of them, while you might not understand their language, you'd probably understand their math. More people "speak math" than any other language in the world. In fact, math is so universal, many people believe that if aliens ever contacted us, they would use the "language" of mathematics.

The "Of" Multiplication Rule

Usually I don't make broad, sweeping statements, but I'm going to make an exception, just this once: When you are doing a word problem and you see the word *of* immediately surrounded by two numbers, then the word *of* stands for "multiply!"*

(first number) **of** (second number) = (first number) × (second number)

Think about it: What is $\frac{1}{2}$ of 12? Multiply $\frac{1}{2}$ times 12, and you get 6. What is 30% *of* 200? We know that 30% = 0.3, so multiply 0.3 times 200, and you get 60.

Remember our milk carton from chapter 13? If a milk carton says 2% milk fat, and you have a half liter of milk, how much *total milk fat* is there in the milk?

Let's see. The question we need to answer here is: "What is 2% of a half liter?" or 2% of 0.5 liter = ?

2% of 0.5 liter = 0.02 × 0.5 = 0.01 liter

That seems like a pretty small amount of milkfat, doesn't it?

QUICK NOTE! For the "of" rule to work, the word *of* must be *immediately* surrounded by two numbers. If it isn't—if there is *anything else* stuck in between the numbers and *of*—there will be an extra step (or steps) involved.

For example: "Find 40% off of $20."

In this case, you should *not* multiply 40% times 20 to get the answer. If you did, you would end up with the wrong answer, because the word "off" is *in between* "of" and one of the numbers!

....

* By the way, the first number is usually a fraction, decimal, or percentage.

Now for another example. Let's find 25% of $20. First we need to convert the 25% to a decimal or a fraction—then we can multiply it by 20. Well, 25% = $\frac{1}{4}$, so we can multiply $\frac{1}{4} \times 20$, and we get 5. Or we could have converted 25% to a decimal and then multiplied 0.25×20. Either way, we get the same answer: $5!

 Doing the Math

Translate these English fragments or sentences into math. (Don't solve them, just practice *translating* them.) I'll do the first one for you.

1. What is 20% of 8?

Answer: 0.2×8

2. What is 0.6 of 10?

3. One third of 30 oranges?

4. What's sixteen percent of $\frac{1}{3}$ of 600?

5. What is 60% off of ten dollars?

(Answers on p. 292)

This last one is sort of a trick question, but I wanted to show you how easy it is to make this particular mistake. The problem said "60% *off* of ten dollars," not "60% of ten dollars." The *of* was not immediately surrounded by two numbers, so you can't just blindly substitute multiplication and get your answer—there is an extra step involved.

What does "60% off" mean? It means that you want to subtract 60% of $10 *off* of the original amount ($10).

$$60\% \text{ of } 10 = 0.6 \times 10 = 6$$

Now we know that we can take $6 "off" of the original $10.

$$10 - 6 = 4$$

Answer: 60% off of $10 is $4.

With this method, it's very important to remember the extra subtraction step. (A common mistake is to think that once you're done with the multiplication part you're *done*—but you're not!) Let's do some more of these.

"Percent Off" Sales

Let's figure out how much our sale item costs. It's always nice to know how much things are going to cost before you get to the register—and it's not hard. It just takes a little math!

Step-By-Step

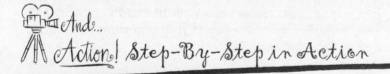

Solving "percent off" problems:

Step 1. Convert the percent to either a fraction or a decimal.

Step 2. Multiply it times the dollar amount, to get a smaller dollar amount.

Step 3. Subtract this "smaller amount" from the *original* dollar amount. Done!

And... Action! Step-By-Step in Action

You've finally found jeans that fit you perfectly. They originally cost $50, but they've been marked down 30%. How much do they cost now?

Translating this word problem into math, it looks like we need to find "30% off of $50."

Step 1. Convert 30% to a fraction or decimal. Let's do a decimal: 30% = 0.30 = 0.3.

Step 2. Multiply it by the dollar amount: $0.3 \times \$50 = \$15$.

Step 3. Subtract $15 (which is 30% of $50) from the original amount: $\$50 - \$15 = \$35$.

Great jeans for $35? Not bad!

"Fraction Off" Sales

Sometimes you'll be asked to find a *"fraction* off of" something. This works the same way as finding a *"percent* off of" something—although it's even easier, because you don't have to convert anything—you can just skip to step 2. Check it out.

What is $\frac{1}{4}$ *off* of $12?

First we find $\frac{1}{4}$ of 12 (which is $\frac{1}{4} \times 12 = 3$). Then we subtract $3 *off* of $12, to get $9. Done!

What Do You Have to Say?

"*I* didn't like math last year, because I didn't like my teacher. I like it now, though—my teacher makes it fun. I like using my head to make problems work." Paolina, 13

"*D*on't be afraid of things you don't understand." C. C., 17

Doing the Math

Solve these "percent off" and "fraction off" problems. I'll do the first one for you.

1. You saw the cutest top for $140, but that's way more than you're willing to spend! There's a sale going on, though, and it's been marked down—75% off! How much does the top cost now?

<u>Working out the solution</u>: We need to find 75% off of $140. First let's find 75% of $140. That would be 0.75 × $140 = $105. Now we need to subtract $105 from $140, since we want 75% off of $140. $140 − $105 = $35. Wow! Sounds like a good buy to me!

<u>Answer</u>: $35

2. You found a great handbag for $30, and it's been marked down—30% off! What's the sale price?

3. There's a sale going on at your favorite shoe store, and the sign says, "Take $\frac{1}{3}$ off of everything in the store!" You find a great pair of ankle boots for $120. How much will they cost now?

4. A magazine subscription says that you'll save 40% off of the cover price each month when you buy a full-year subscription. The normal cover price is $2.50. How much will you pay for each magazine now? How much will you pay for the full-year (12 month) subscription?

(Answers on p. 292)

There will be other types of word problems in the following chapters, so take a look at this list of handy translations!

Other Helpful "Translations" from English to Math*

English	Math
of	× (multiplication)
sum, total, more than	+ (addition)
difference, less than	− (subtraction)
per, quotient, a	÷ (division)
is, are	= (equal sign)

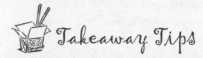 Takeaway Tips

- Think of word problems as English that needs to get *translated* into math. Math is just another language, after all!

- When you see *of* in a word problem, and it's immediately surrounded by two numbers, you can automatically substitute a multiplication sign for the *of*.

- If you need to find a *"percent off of"* a price, just find that "percent of," and subtract it from the total.

- If you need to find a *"fraction off of"* a price, just find that "fraction of," then subtract it from the total.

Do You Speak "Math"?

*J*f you think about it, math really is a foreign language. It primarily uses numbers (and sometimes letters like *x* and *y*) to express ideas—not the "regular" English that we're all used to. Even other sciences like chemistry and physics use more "regular words" than math does.

. . . .

* *For division, if **a** can be replaced by the word **per** and not change the meaning of the sentence, then **a** translates into "divide by."*

Math is the language that your computer uses to communicate internally, the language that dictates chemical formulas for everything from plastics to lipstick, and the language your cell phone uses to send text messages through the air. It's often called the "language of the sciences." It's a special language, like a secret code, and if you can speak "math," you can understand things other people can't.

Of course, the downside is that it's often hard to ask other people, who may not speak "math," for help. And just like a foreign language, if you don't practice it, it's easy to forget it. That's why math is always harder after summer break, and why most math classes spend the first month or two reviewing what was taught the year before.

If you think of math as a foreign language, it may make more sense why it can sometimes seem confusing. Be patient! After all, would you sit down on your first day of Spanish class expecting to already understand Spanish? Of course not. You'd expect that it might not make much sense at first, but if you stick with it, slowly but surely, it will become clearer and easier to understand.

The same is true of the language of math, and this book will help you speak it much more fluently.

Chapter 16

Does She Ever Get Off the Phone?

Ratios

$\mathcal{W}$hen I was in high school, I used to talk on the phone after school for *hours* at a time, mostly to my best friend, Kimmie.

Sure, I spent a lot of time on my homework, too, but unless I had a test the next day, I would say that for every *1 hour* I spent on homework, I spent *2 hours* talking on the phone.

That's a **ratio** of 1 to 2.

There are three ways of writing *ratios*:

$$1 \text{ to } 2 \quad 1{:}2 \quad \frac{1}{2}$$

All three ways of expressing this ratio do the same thing: they compare 1 to 2.

Ratios can compare hours, miles, money—any kind of units you want, so long as both of the numbers you are comparing have the *same units* as each other. In other words, when working with ratios, you'll always compare hours to hours, miles to miles, and so on. Never hours to miles. See what I mean?

You can think of ratios as a "real-life" way of *using* fractions—and they come up in word problems, too, so they are good to know! We've spent several chapters learning how to manipulate the fractions themselves, and now we'll start seeing what we can use them *for*. (There must be more uses than just pizza!)

What's It Called?

Ratio

A *ratio* is just a comparison of two numbers or values, expressed in the same terms, or *units*. In other words, in ratios we always compare apples to apples or hours to hours, and so on. A ratio can be written in words, with a colon, or as a fraction. Ratios are *always* written in reduced form.

Sure I talked on the phone with Kimmie a lot, but on a night before a *test*, I wouldn't spend as much time on the phone. For every 1 hour I spent studying, I probably only talked to Kimmie for 20 minutes.

Hmm, 1 hour compared with 20 minutes. What would that ratio of "study hours to phone hours" look like?

$$1 \text{ to } 20? \quad 1:20? \quad \tfrac{1}{20}?$$

Nope—that would mean that for every hour I spent studying, I talked on the phone for 20 hours! Oops.

What went wrong? The problem was: we were trying to compare two different *units*—hours and minutes—and the *units* always need to be <u>the same</u> before we can make a ratio.

QUICK NOTE! Whether you are writing a ratio with a colon or as a fraction, the two numbers should have no common *factors*; in other words, a ratio should always be in *reduced* form.

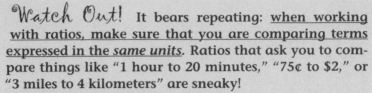

Watch Out! It bears repeating: <u>when working with ratios, make sure that you are comparing terms expressed in the *same units*.</u> Ratios that ask you to compare things like "1 hour to 20 minutes," "75¢ to $2," or "3 miles to 4 kilometers" are sneaky!

Why are they sneaky? Well, hours and minutes are both units of *time*, dollars and cents are both units of *money*, and miles and kilometers are both units of *distance*. But when the units are slightly different like this, you can get into trouble, like we almost did with the "1:20 hours" ratio above.

When the units are different, you must do a **unit conversion** to get them to be the same. We'll learn more about unit conversions in chapter 19. It's basically what it sounds like, though—converting from one unit to another. For now, we'll keep the unit conversions very simple. Like changing dollars to cents or changing hours to minutes.

Just stay alert, and you won't be fooled!

Keeping all of this in mind, if we want to **find the ratio of 1 hour to 20 minutes**, we should express both either in minutes *or* hours.

Let's convert the 1 hour into minutes. We know that 1 hour = 60 minutes, so, substituting that in, the ratio would be 60:20 or $\frac{60}{20}$.

Yes, but that's not the final answer, since ratios must always be expressed in their simplest, most reduced form. The final step is to reduce this fraction—however you'd like! You can make lists of factors (or do the Birthday Cake method) and discover that the GCF is 20, then divide the top and bottom by 20—or you can hack away at the numbers until it's reduced.

$$\frac{60}{20} = \frac{60 \div \mathbf{20}}{20 \div \mathbf{20}} = \frac{3}{1}$$

So, the answer is: on a night before a test, my ratio of study hours to phone hours was 3 to 1 or 3 : 1 or $\frac{3}{1}$. Done!

QUICK NOTE! If you are given two different units in a ratio, and you need to convert one to the other, you can pick *either* one. Either choice will result in the same ratio, once you're done reducing. Whichever unit looks easier to deal with, that's the one to use!

Watch Out! Watch the order of your numbers. If the problem asks for "study hours to phone hours," make sure you put the study hours first, and on top of the fraction. (This makes sense, but it's a common mistake!)

Step-By-Step

Finding ratios:

Step 1. Make sure both values are expressed in the *same units*. (If not, then do a quick *unit conversion*.)

Step 2. Create a fraction from the two numbers. The *first* number should go on *top*. In other words, the ratio of "this to that" $= \frac{this}{that}$

Step 3. Reduce the fraction, and you've got your ratio! You can leave the final answer as a fraction, put it in words, or just use a colon to separate the two numbers.

And... Action! Step-By-Step in Action

Between your backpack, your purse, and your bathroom drawer, you've got 15 different kinds of lip gloss! Sure, some of them are just collecting dust—but your little sister "only" has 12 different kinds, and yes, she's jealous of you. What's the ratio of how many lip glosses *you* have compared to how many *your sister* has?

Ratio of lip glosses to lip glosses = ?

Step 1. Are the 15 and 12 expressed in the same units? Let's see—both are numbers of lip glosses. Yep!

Step 2. Create the fraction: $\frac{\# \text{ of your lip glosses}}{\# \text{ of your sister's lip glosses}} = \frac{15}{12}$.

Step 3. Reduce it: $\frac{15}{12} = \frac{15 \div 3}{12 \div 3} = \frac{5}{4}$. Done!

The ratio of your lip glosses to hers is $\frac{5}{4}$. You could also write the answer as 5 to 4, or 5:4.

This means that for every 5 lip glosses you have, she only has 4. But I'm sure you let her borrow yours, sometimes . . . right?

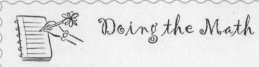
Doing the Math

Find the following ratios, and be sure to leave them in reduced form. I'll do the first one for you.

1. I spent $1.25 on a cookie and 50¢ on an apple. What's the ratio of money spent on the cookie to money spent on the apple?

<u>Working out the solution</u>: First, make sure they are in the same unit. Let's change $1.25 to 125¢, so that both amounts are in "cents." Now, notice that it asks for "money spent on the cookie" *first*, so we'll compare 125¢ to 50¢ in a fraction, putting the first one on top: $\frac{125}{50}$, and reduce. Let's see—25 divides into the top and the bottom, so we get: $\frac{125}{50} = \frac{125 \div 25}{50 \div 25} = \frac{5}{2}$.

<u>Answer</u>: The ratio of money spent on the cookie to the apple is "5 to 2," 5:2, or $\frac{5}{2}$.

2. You texted your boyfriend 16 times today, and he texted you 12 times. What's the ratio of your texts to his?

3. You were so hungry yesterday after swimming that you ate $1\frac{1}{2}$ sandwiches. Your best friend, who didn't swim, ate just $\frac{1}{2}$ of a sandwich. What's the ratio of how much you ate, compared to how much your best friend ate? (Hint: Make a big complex fraction and then simplify to get the ratio.)

4. You went to the store and spent $6 on makeup and 60¢ on a banana. What's the ratio of money spent on makeup compared to money spent on the banana?

5. A pack of gum has 5 individual *pieces* of gum in it. You have 2 *packs* of gum, and your friend has 6 individual *pieces* of gum. What's the ratio of how much gum you have, compared to your friend? (Hint: You'll need to do a quick unit conversion—you'll want to write the ratio in terms of pieces of gum, not packs.)

(Answers on p. 292)

What's the Deal?

You may have noticed that in the solution to the "phone hours to study hours" ratio on page 196, what we really found was the ratio of "phone *minutes* to study *minutes*." After all, we converted the times into *minutes*—20 and 60—and then reduced.

Knowing this, you might be thinking, "Why did we say 'study *hours* to phone *hours*' in the answer, instead of 'study *minutes* to phone *minutes*?'"

As it turns out, for ratios, they're really saying the same thing!

Think about it: if for every minute you spend on the phone, you spend 3 minutes studying, then for every hour you spend talking on the phone, you're spending 3 hours studying, too. Does that make sense? The only thing that matters in ratios is the relationship *between* the two numbers—not the numbers themselves.

But to illustrate how they are equivalent, we also could have done this problem by converting everything to *hours*, and then simplifying *that* ratio instead. If you want to see this done, check out this book's website: mathdoesntsuck.com.

" '*Dumb*' girls are smart girls who act dumb for guys—but they have all the potential of 'smart' girls!" Brianna, 17

"*I* don't believe that truly dumb girls exist. I think girls who are labeled as dumb are just the ones who have allowed the stereotypes that society creates to define who they are." Iris, 15

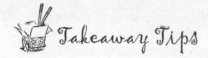

Takeaway Tips

- When making a ratio from two numbers, make sure both numbers are expressed in the *same exact units* (i.e., make sure you're comparing hours to hours, miles to miles, lip glosses to lip glosses, and so on).

- Once both units are the same, make the two numbers into a fraction, reduce it, and you're done!

TESTIMONIAL:

Tricia Hacioglu (New York, NY)

<u>Before</u>: Disastrous cookie baker!
<u>Today</u>: Foreign-exchange trader at an
investment bank

Hi! My name is Tricia. I'm a foreign-exchange trader
at an investment bank. My job is to estimate how
the value of one currency will change compared to
another—in other words, whether one country's money
(say, the euro) will become more or less valuable
relative to another country's money (say, the
Japanese yen); then I trade the "exchange rate" in
order to make a profit. It's kind of like trading
stocks, where the stocks themselves are money from
different countries.

My job is exciting because I have to figure out how
things that happen in the news are going to affect
the currency rates—and believe me, they really do!

> "Life is so
> much better
> when you're
> equipped
> to do a job
> you love."

Sometimes, the currency values change
within *one minute* of a story (say, an
earthquake in Italy or a large company
going bankrupt in Japan) hitting the
news. In order to make these trading
decisions, I must constantly use logic
and math. Then, at the end of each day,
I calculate my daily profits and losses
(hoping for lots of profits and no
losses!) to see how I traded. It's fast-
paced and challenging, and I love it.

Life is so much better when you're *equipped* to do
a job you love—even if you didn't even know that job
existed when you were first learning math.

COOKIE BAKING 101: A DISASTER WAITING TO HAPPEN!

When I was in elementary school, I loved baking
cookies. And it wasn't long before I began to get

creative with the recipes. "How would these taste with more sugar and oil?" I'd wonder. "I bet extra butter will make these especially crispy." Unfortunately, my cookies weren't turning out very tasty, because I wasn't adjusting the various ingredients proportionately.

It was then that I realized the importance of being able to perform simple mathematical functions (like fractions and proportions) in everyday life. These skills helped me get my cookie recipes back on track—and are now essential for my job!

Even if you don't know how or when you will ever use what you're learning in math class in the real world, believe me—someday, you will. Math skills come in handy in so many situations in life. Hang in there, and you'll be grateful you did!

Chapter 17

The Perks of a Southern Drawl

Rates and Unit Rates

*M**agazine subscription savings! Special Internet* **rate** *of only $2.99 per issue!*

With about 820,000 pregnant teenage girls per year, the United States has the highest teen pregnancy **rate** *in the western world . . .**

Get your heart **rate** *up to 120 beats per minute with aerobic exercise! Check the chart to see how many beats per minute is best for you!*

Rates are everywhere. A *rate* is like a ratio, except that it compares two quantities that have *different* units.

....
* *As of May 2005. Source: www.teenpregnancy.org/resources/data/genlfact.asp*

A *rate* is a comparison of two quantities with *different* units. Rates are usually written as fractions and are always expressed in simplest, reduced form.

Since rates involve two different units, you must always <u>include the units</u> as part of the fraction, and in your final answer.

For example, if you and 10 friends had an ice cream party with 2 buckets of ice cream (that's you plus 10 friends = 11 people), the *rate* of ice cream to people is:

$$\frac{2 \text{ buckets of ice cream}}{11 \text{ people}}$$

Since 2 and 11 have no common factors, the rate is in simplest form and doesn't need to be reduced. Done!

The most popular kind of rate is called a **unit rate**. In a unit rate, <u>the denominator is always 1</u>. All the quotes from the beginning of the chapter are *unit rates:*

$$\frac{2.99 \text{ dollars}}{1 \text{ magazine}} \qquad \frac{120 \text{ heartbeats}}{1 \text{ minute}} \qquad \frac{820{,}000 \text{ pregnant girls}}{1 \text{ year}}$$

Often with unit rates, you'll see the word *per* used: "dollars per magazine," "heartbeats per minute," and so on.

What's It Called?

Unit Rate

A *unit rate* is a rate in which the denominator (bottom) is equal to 1. Unit rates can be expressed as fractions (see above), or using the word *per*. "65 miles per hour," "2 hamburgers per person," "17 songs per CD," etc.

If the rate involves money, like "$20 per DVD," then the unit rate is called the **unit price**. Makes sense, right?

QUICK NOTE! The numerator (top) in a unit rate doesn't have to be a whole number—in fact it often isn't. (Like the magazine example at the beginning of the chapter.) Another example: if you drink 1.5 liters of water a day, then your *unit rate* of drinking water would be 1.5 liters per day, in other words $\frac{1.5 \text{ liters}}{1 \text{ day}}$. Unit rates are a special case where you don't reduce/simplify the fraction in the typical way. (Remember that in unit rates, the denominator must always = 1.)

Step-By-Step

Writing rates and unit rates:

Step 1. Write the rate as a fraction. If the question asks, "What is this *to* that" or "this *per* that," then the fraction should look like: $\frac{this}{that}$. And always include the units in the fraction, since you'll have 2 different units.

Step 2. Reduce the fraction—unless you are doing a *unit rate*, in which case you can skip to step 3.

Step 3. For *unit rates*, divide the top and bottom *by the denominator*, so you get a 1 as the new denominator. (Remember that dividing top and bottom by the *same number* will have no effect on the value of the fraction). Note: The numerator doesn't need to be a whole number—it might be a decimal or even a fraction.

Step 4. Don't forget to include the units in your answer!

QUICK NOTE! When writing a rate, whatever appears right after the word *per* will always be the denominator.

Sometimes, questions will be worded something like this: "12 puppies ate 6 pounds of dog food. Per puppy, how many pounds of dog food were eaten?" When this is the case, how do you know what to put on the top and bottom of the fraction?

Since we see the word *per*, all we have to do is look for the word immediately *following* "per"—in the example above, it's "puppy"—and stick that on the bottom.

So, our answer would be $\frac{6 \text{ lb.}}{12 \text{ puppies}} = \frac{6 \div 12}{12 \div 12} = \frac{0.5 \text{ lb.}}{1 \text{ puppy}}$ or "0.5 lb. of dog food per puppy." Each puppy ate 0.5 lb. of dog food.

You now know that the word *after* "per" is the denominator. But how are you going to remember the order? Easy: Think of the word *pretty*. Now imagine that you are saying it with a Southern drawl, and it sounds more like "perdy." Per-D. Get it? The thing that comes right after the *per* is D, the denominator!

And... Action! Step-By-Step in Action

Write the following scenario as a rate in reduced form.

For the class field trip, the bus broke down and the teachers had to drive the kids in their cars! There were 30 kids and 8 cars. What was the rate of kids to cars?

Step 1. Write it as a fraction: $\frac{30 \text{ kids}}{8 \text{ cars}}$

Step 2. Reduce the fraction. Well, 2 goes into both 30 and 8, so we get: $\frac{30 \div 2}{8 \div 2} = \frac{15}{4}$. And 15 and 4 don't share any common factors, so it's reduced!

Step 3. If we were doing a unit rate, we'd get $\frac{15 \div 4}{4 \div 4} = \frac{3.75}{1}$, or 3.75 kids per car. Oops, you can't really have 3.75 kids now, can you? No, not even your annoying little brother should be considered 0.75 of a kid. So a unit rate doesn't really apply in this case. We'd better stick with the reduced fraction rate: $\frac{15}{4}$.

Step 4. Don't forget to include the units! Answer: $\frac{15 \text{ kids}}{4 \text{ cars}}$

Take Two! Another Example!

Your favorite store is having a sale on some really cute T-shirts. They normally cost $4 each, but during the sale you can get 5 shirts for $18. What's the price per shirt if you buy 5 of them? In other words, if you buy 5 shirts during this sale, what unit price will you be paying?

Step 1. Write out the fraction. Since it says "per shirt" (see the "Quick Note" on p. 205), we know that *shirts* goes in the denominator: $\frac{\$18}{5 \text{ shirts}}$ (Remember: Per-*D*!)

Step 2. Since it's asking for the *unit price*, we can skip to step 3.

Step 3. Divide top and bottom by 5 to get a 1 in the denominator: $\frac{18 \div 5}{5 \div 5}$. The top now requires a little decimal division (see p. 125 for a review): $18 \div 5 = 5\overline{)18} = 3.6$. So our new unit fraction is $\frac{3.6}{1}$.

Step 4. Don't forget to include the units! Answer: $\frac{\$3.60}{1 \text{ shirt}}$ or "$3.60 per shirt."

So it's a better deal than $4 a shirt, but not *that* much better of a deal.

Doing the Math

Write these situations as *rates*, and make them *unit rates* unless your answer doesn't make sense (like with the 3.75 kids!). I'll do the first one for you.

1. Your mom bought some hamburger meat at the store. You notice that the sticker says $6.72 and that it weighs 2.1 lbs. How much did the meat cost per pound?

<u>Working out the solution</u>: First write the rate as a fraction. Since it's asking for "cost per pound," we know that pounds goes on the bottom: $\frac{\$6.72}{2.1 \text{ lbs.}}$.

Because money is usually expressed in decimals, we know we can do a unit rate. We might as well skip to step 3 and divide the top and bottom by the denominator to get a 1 on the bottom. So, $\frac{6.72 \div 2.1}{2.1 \div 2.1}$. The bottom will be 1, and now let's do the top.

We need to do decimal division to solve $6.72 \div 2.1 =$ $2.1\overline{)6.72} = 21\overline{)67.2}$ (Because we had to get rid of the decimal point in 2.1, we multiplied both by 10).

A little decimal division (or a calculator) will tell us that $21\overline{)67.2} = 3.2$. So, reviewing all the steps we just did:

$$\frac{\$6.72}{2.1 \text{ lbs.}} = \frac{6.72 \div \mathbf{2.1}}{2.1 \div \mathbf{2.1}} = \frac{3.2}{1}$$

And since the 3.2 is *dollars*, we know that 3.2 = $3.20. *Always* include the units in a rate fraction.

<u>Answer</u>: the hamburger meat cost $\frac{\$3.20}{1\text{ lb.}}$ or $3.20 per pound. And if you went to the grocery store, you'd probably see that price in the window!

2. You've been training for an upcoming track meet, and your coach says you've run a total of 32 miles in 5 days. Per day, how many miles is that?

3. Yesterday, 10 kids participated in the school's fund-raising car wash; they washed 14 cars in all. What was the rate of kids to cars?

4. You bought 5 bottled waters for you and your friends. The total bill was $12.50. How much did it cost per bottle? In other words, what is the unit price?

5. You bought some ribbon at the store to decorate your photo album. You bought 3.2 feet of ribbon, and it cost $2.88 total. What was the unit price per foot?

(Answers on p. 292)

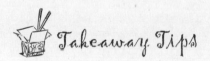 Takeaway Tips

- *Rates* compare two numbers with different units—always be sure to include the units in your answer.

- *Unit rates* are special rates with a denominator of 1. (The numerator can be a decimal or even a fraction itself!) These types of rates are used often in real life.

- When you're first writing a rate fraction, make sure to *put the correct amount on top and bottom*. For example, if it asks, "How many cars *to* students are there?" then you put cars on top and students on the bottom. If the word *per* is in the question then you're golden: whatever quantity comes immediately *after* the word *per* will always be the *denominator*. (Just think Per-*D*!)

Let me tell you something: Knowing how to use rates can help you with more than just your math homework. Believe it or not, I once used unit rates to help me look my best for a movie shoot!

Danica's Diary

FACE CREAM FIASCO

A few years ago, I traveled all the way to the Eastern European country of Bulgaria to star in a television movie called *Path of Destruction*. Before I left, I was packing my bag, trying to think of everything I would need for the whole five weeks that I would be gone.

You never know what products you will be able to buy once you're in a foreign country, so you have to pack well! Did you know that you simply cannot find peanut butter in Bulgaria? It's true. My mom came to visit me while I was there, and when she flew through Paris, she picked up some peanut butter for me on her way! Now that's a good mom.

Anyway, I was pretty sure that if I ran out of certain items, like toothpaste, I could just buy whatever the local stores were selling. But there were a few items I did not want to take my chances on. One of those items was a face moisturizer that I just *love*. I love how it feels, and it helps keep me from breaking out, which is a huge plus.

I wanted to make sure I brought enough with me for all 5 weeks. But I didn't really know how much I would need, because I had never really paid attention to how much I used per week! I mean, who pays attention to that sort of thing?

I needed to figure out: At what *rate* did I use the lotion? How much per day? Per week?

What if I ran out? I knew that I could always just bring 3 full jars with me, but it's kind of expensive moisturizer, and I didn't have much extra space in my bags, and I *really* wanted to bring my curling iron—so I only wanted to bring as much as I would actually need.

A few days before my trip, I still had no idea how I was going to figure it out. Then, one afternoon, I was standing in the bathroom, staring in despair at my almost-empty 2-ounce jar of lotion, when I suddenly remembered that I also had a tiny "free sample" packet of the moisturizer.

I searched through my drawers and found it—and saw that it was labeled 0.14 oz. Aha! I could use that free sample to determine the *unit rate* of lotion that I use per *day*. Then, once I knew how much I used per *day*, I could figure out how much I used per *week*. Here's how I did it.

To determine the *unit rate* of use, I would need to plug in the amount of lotion I used and the number of days that I used it in.

$$\frac{\text{ounces of lotion}}{\text{days}}$$

So, for the next few days, I used lotion from the free sample and learned that it took me just 2 days (using it in the morning and at night) to use up the 0.14 oz. packet. I was then able to express this as a *rate:* $\frac{0.14 \text{ oz.}}{2 \text{ days}}$

Next, I converted it into a *unit rate*, by dividing the top and bottom by 2.

$$\frac{0.14 \text{ oz.}}{2 \text{ days}} = \frac{0.14 \div 2}{2 \div 2} = (\text{use decimal division}) = \frac{0.07 \text{ oz.}}{1 \text{ day}}$$

I learned that I used 0.07 oz. of the moisturizer per day. Now I just needed to multiply 0.07 oz. times the number of days I'd be in Bulgaria to figure out how much moisturizer I would use while I was there.

Since there are 7 days in one week, in 5 weeks there are $7 \times 5 = 35$ days. And if I needed 0.07 oz. for just 1 day, how much would I need for 35 days? This part was easy: all I needed to do was multiply 0.07 oz. by 35.

$$0.07 \times 35 = 2.45 \text{ oz.}$$

So I concluded that I'd need 2.45 oz. for the entire trip; that's almost $2\frac{1}{2}$ oz. Since 1 jar only holds 2 oz., I knew that if I brought 2 jars with me, I'd be totally safe.

My trip was great, I had an awesome time shooting the movie, and thanks to a little math, I didn't have to worry about running out of my favorite moisturizer.

Chapter 18

Filmmaker Extraordinaire!

Proportions

$\mathcal{S}$arah and Madison are complete opposites. Sarah loves unicorns and mermaids, and Madison loves hard rock music and black nail polish. But they do have one thing in common: they both have younger sisters (who love to imitate their older sisters). Let's say that Sarah's younger sister's name is Sue, and Madison's younger sister's name is Meg. Then we could say:

$$\frac{Sarah}{Sue} = \frac{Madison}{Meg}$$

Even though Sarah and Madison are so different, the *relationship* between Sarah and Sue is the same as the *relationship* between Madison and Meg—which is what the **proportion** above expresses.

Proportions are really handy, and they're pretty intuitive. Have you ever heard of comparisons like "gloves *are to* hands *as* socks *are to* feet?"

We're saying that the *relationship* between gloves and hands is the same as the *relationship* between socks and feet. (Especially those socks with individual toes. Have you ever seen those? They're kinda creepy, but fun at the same time.)

$$\frac{\text{gloves}}{\text{hands}} = \frac{\text{socks}}{\text{feet}}$$

And in this case, each thing in the numerator is *worn* on the thing in the denominator. Gloves are *worn* on hands, and socks are *worn* on feet.

How about this one: "<u>oak</u> *is to* <u>tree</u> *as* <u>tulip</u> *is to* <u>flower</u>"? You could express the comparisons above like this:

$$\frac{\text{oak}}{\text{tree}} = \frac{\text{tulip}}{\text{flower}}$$

In this case, each thing in the numerator is a *type* of the thing in the denominator. Oak is a *type* of tree, and tulip is a *type* of flower.

What both of these comparisons communicate is that the relationship between the two items on the left is *equivalent* to the relationship between the two items on the right. Yes, kind of like equivalent fractions.

Again, it's not the items themselves, but the *relationship between them* that matters. Do you know what goes in the missing blank here?

$$\frac{\text{fingernail}}{\text{finger}} = \frac{?}{\text{toe}} *$$

In life, these kinds of comparisons are called *analogies*. In math, they are called *proportions*. And while some of these examples may seem silly and obvious, believe me, thinking about these kinds of comparisons will help you to better understand proportions—the focus of this chapter.

For instance: In the proportion $\frac{1}{5} = \frac{10}{50}$, each thing on the bottom is *five times as big* as the thing on top. Both ratios have the same *relationship* to each other; they are equivalent fractions.

What's It Called?

Proportion

A *proportion* is a statement showing the equality between two fractions. The fractions are usually ratios or rates. Since the fractions in proportions are equivalent, the <u>cross products will always be equal as well</u>. (See p. 79 for a review of cross multiplication.)

• • • • • • • • • • • • • • • • • • •

* *Answer: Toenail. Yes, I know. This is super-complicated stuff.*

Danica's Diary

MOVIE MAGIC

When I was 12 years old, I got my first home video camera, and I used to love making short little movies with my sister, Crystal, and my cousin, Elena.

Mostly we did "remakes" of movies that we liked: *Superman*, *Batman*, *Cinderella*, *Frosty the Snowman* (we often got together around the holidays), and so on. But let's face it: we only had stuff around the house to use as props, and there were only 3 of us, so we had to use our imaginations quite a bit.

In our remake of *Cinderella*, Crystal (the only blond) played Cinderella, my cousin Elena played the evil stepmother, and I played *both* stepsisters—so I had to keep changing my clothes back and forth. In our remake of *Frosty the Snowman*, when Frosty melted, I got a shot of Elena sinking to her knees, then got a reaction shot of Crystal crying and saying, "Nooo! Don't melt," then cut back to a shot of the melted snowman, which amounted to nothing more than a black hat and a little cup of water just sitting on the floor.

Okay, they weren't exactly professional, but everyone's gotta start somewhere.

Are you an aspiring filmmaker? You probably have "movie mode" on your digital camera, right? Let's say we wanted to make a little horror movie where—let's see—a giant monster dog invades a small village.

What sort of props would we need? Well, we'd definitely need some small people to play the townspeople. We all keep some Barbie and Ken dolls in a box somewhere in the attic, right? Barbies might be a thing of the past, but they're helpful for those of us who want to practice filmmaking. As long as you say their lines for them offscreen, Barbies can be very poised in front of the camera.

What else would we need? Well, your dog Sparky could come in handy!

So, it's decided: Sparky will play the part of Monster Dog, and Barbie and Ken will play the townspeople. Let's call it *Attack of the 50-foot Monster Dog*.

Then again—I'm not really sure if Sparky would *look* 50 feet high when compared to Barbie and Ken. So maybe that's not the best title. Should we call it *Attack of the 10-foot Monster Dog? Attack of the 100-foot Monster Dog?* I wonder how tall Sparky will *seem* in the movie, when compared to our Barbie and Ken townspeople?

Let's take a moment and organize our thoughts:

- Ken's real height is about 1-foot tall.
- Sparky's real height is 3-feet tall. (He's a Great Dane—a big guy!)
- Ken should *seem* about 6-feet tall in the movie, since that's approximately how tall he would be if he were an actual person.
- Okay. So if Ken seems 6-feet tall in the movie, how tall will Sparky seem?

This is a perfect situation to use *proportions*.

Ken's fraction   Sparky's fraction

$$\frac{\text{real height}}{\text{movie height}} = \frac{\text{real height}}{\text{movie height}}$$

We can express this as two equivalent fractions.

Ken   Sparky

real life → $\dfrac{1 \text{ foot}}{6 \text{ ft.}}$ $=$ $\dfrac{3 \text{ ft.}}{?}$ ← real life

seems in movie →   ← seems in movie

On the left-side fraction we have Ken's info: In real life, Ken is 1-foot tall, but he'll seem 6-feet tall, since he's supposed to be a human being.

On the right side, we have Sparky's info: Sparky is 3-feet tall in real life, and the "?" is the spot where we'll eventually plug in the height that Sparky will *seem* to be in the movie. But how do we get that missing number?

Well, it's a proportion, which means that the relationship between the top and the bottom of the fraction must be the same in each fraction. For Ken, the bottom of the fraction is 6 times as big as the top. That must also be true for Sparky. So, what is 6 times as big as 3 feet? Well, 6 × 3 ft. = 18 ft.

Ken   Sparky

$$\frac{1 \text{ ft.}}{6 \text{ ft.}} = \frac{3 \text{ ft.}}{18 \text{ ft.}}$$

We're almost done. Before we finish, though, let's check the cross products to make sure we have a real proportion:

$$\frac{1}{6} \overset{18 \quad\; 18}{\underset{}{\times}} \frac{3}{18}$$

Yep, 6 × 3 = 18 and 18 × 1 = 18, so the cross products are equal, and we have found the right "movie height" for Sparky!

This proportion tells us that Sparky would look 18-feet tall in the movie. So maybe the best title would be *Attack of the 20-foot Monster Dog*. After all, in moviemaking it's okay to exaggerate a *little* bit.

As you saw in this example, sometimes we can find the missing number in a proportion simply by figuring out what we can multiply or divide by to make the fractions equivalent. Other times, however, there is no number that does that, so here's another way to do it! (And personally, I prefer the method below for all of these "find the missing number" problems.)

Using Cross Multiplication to Solve for the Missing Number

What if we had a proportion like this to solve?

$$\frac{4 \text{ ft.}}{6 \text{ ft.}} = \frac{10 \text{ ft.}}{? \text{ ft.}}$$

You could think to yourself, "Four times *what* equals 10? What-ever that number is, I'll have to multiply the *same number* times 6, to get the answer." But in this case, that number would end up being a fraction...yikes!

You could figure out what fraction does that, or you could use *cross multiplication* to figure out what the missing value is supposed to be. Even though there's still a "?" in one of the spaces, let's cross multiply and set them equal to each other. We'll just keep the "?" where the missing number should go:

$$\frac{4}{6} \diagdown\hspace{-0.9em}\diagup \frac{10}{?}$$

We know that in every proportion, the cross products are equal—so we just need to figure out what number goes in the "?" spot to make the cross products, in fact, equal.

If the cross products are equal, that would mean:

$$4 \times ? = 60$$

Do you know how to find the missing "?" now? What needs to multiply times 4, to get 60? That sounds like multiplication in reverse. Yep, we just *divide*: $60 \div 4 = 15$. And now we've found the missing number, because if $60 \div 4 = 15$, then $4 \times 15 = 60$.

We can write the original proportion now, filling in 15 for the missing number "?" and find the cross products to double-check our answer.

$$\frac{4}{6} \diagdown\hspace{-0.9em}\diagup \frac{10}{15}$$

So, $6 \times 10 = 60$ and $15 \times 4 = 60$. Yep! We have equivalent fractions, so we've found the missing number in the proportion.

Let's do another one.

Find the missing number in this proportion:

$$\frac{4}{5} = \frac{?}{30}$$

Instead of calling it "?" this time, we'll call the missing number *m* (for <u>missing number</u>). And let's find the *cross products* and set them equal to each other, because in a proportion the cross products *have* to be equal, even if we don't know what *m* is yet:

$$\frac{4}{5} \diagdown\hspace{-0.9em}\diagup \frac{m}{30}$$

If the cross products are equal, that would mean:

$$120 = 5 \times m$$

What should m be equal to? Well, we just need to divide 120 by 5, right? So, $120 \div 5 = 24$. And we have found that $m = 24$!

Let's check our work by doing cross multiplication on our filled-in proportion:

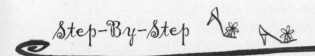

Yep, the cross products are equal: $120 = 120$. So we've found the right number! $m = 24$.

Step-By-Step 👠 👠

Proportions: using cross multiplication to find the missing number:

Step 1. Find the cross products for the two fractions, using m where the missing number would be.

Step 2. Set the two cross products equal to each other, and multiply out the side that has only numbers in it.

Step 3. Use division to figure out what the missing number is.

Step 4. Once you've found what m equals, rewrite the proportion and double-check that the cross products are equal. If they are equal, you've done it! You've found the m that makes the two fractions a proportion.

The "What If?" Game

Regarding step 2, many students ask, "But how do you *know* that the cross products are going to be equal? How can you just "set" them equal to each other?"

That's a good question, and the answer is: We're really just playing a game of "what if?" *What if* the cross products were equal? *Then* what would m have to be?

When we act "as if" the cross products are equal, and we go ahead and "set" them equal to each other, we can find the m that makes the "what if" actually *true*.

And... Action! Step-By-Step in Action

Find the missing number in this proportion.

$$\frac{12}{3} = \frac{m}{12}$$

Step 1. Find the two cross products:

$$\overset{144}{\underset{3}{12}} \overset{3 \times m}{\underset{12}{m}}$$

Step 2. Set the two cross products equal to each other. Here's where we play the "what if?" game. *What if* these two fractions were a proportion? *What if* the cross products were equal? Then what would *m* have to be, in order to make the "what if" come true?

$$144 = 3 \times m$$

Step 3. So, 3 times *what* equals 144? Let's divide 144 ÷ 3 to find out: 144 ÷ 3 = 48. We can check that 3 × 48 = 144, so *m* = 48.

Step 4. Let's double-check our proportion's cross products to make sure we got the right *m*:

$$\overset{144}{\underset{3}{12}} \overset{144}{\underset{12}{48}}$$

Yep, 144 = 144, so we've made the "what if" true by finding the correct *m* to make the cross products equal: *m* = 48.

Doing the Math

Find the missing number that makes these proportions true. I'll do the first one for you.

1. $\frac{7}{4} = \frac{14}{m}$

<u>Working out the Solution:</u> First, we find the cross products:

$$\overset{7 \times m \qquad 56}{\frac{7}{4} \diagdown\!\!\!\!\diagup \frac{14}{m}}$$

Next, we set the two cross products equal to each other. In other words, "what if" they were equal? So, $7 \times m = 56$. Then what must m be? Since $7 \times 8 = 56$ (or literally doing step 3, since $56 \div 7 = 8$), we know that $m = 8$. We can check our answer by looking at the cross products for our new proportion:

$$\overset{56 \quad ? \quad 56}{\frac{7}{4} \diagdown\!\!\!\!\diagup \frac{14}{8}}$$

Since $7 \times 8 = 56$ and $4 \times 14 = 56$, we found the right m!

<u>Answer:</u> $m = 8$

2. $\frac{8}{3} = \frac{m}{9}$

3. $\frac{8}{6} = \frac{m}{9}$

4. $\frac{1}{2} = \frac{10}{m}$

(Answers on pp. 292–3)

Using Proportions to Solve Word Problems

〜〜〜〜〜〜〜〜〜〜〜〜〜〜〜〜〜〜〜〜〜〜〜

I like to keep my body healthy and in good shape, on the inside and out. And over the years, I've noticed that the more water I drink, the clearer my skin is! I'm not kidding—it's true.

Let's say that for every 2 miles I run, I need to drink 1.2 liters of water to stay hydrated. What if I run 2.5 miles? How much water should I drink then?

Proportionally speaking, we're asking: "2 miles *is* to 1.2 liters of water, *as* 2.5 miles *is* to what?" So our proportions will look like this:

$$\frac{\text{miles}}{\text{liters of water}} = \frac{\text{miles}}{\text{liters of water}}$$

Now let's fill in the numbers we know and use *m* for the missing number of liters that should go in the fraction with the 2.5 miles.

$$\frac{2 \text{ miles}}{1.2 \text{ liters}} = \frac{2.5 \text{ miles}}{m \text{ liters}}$$

Next, let's find the cross products and set them equal to each other, just like we did in the previous section.

$$2 \times m \qquad 1.2 \times 2.5$$
$$\frac{2}{1.2} \diagdown \frac{2.5}{m}$$

Setting the cross products equal to each other:

$$2 \times m = 1.2 \times 2.5$$

Since $1.2 \times 2.5 = 3$ (see p. 122 for decimal multiplication), this makes our equation:

$$2 \times m = 3$$

Hmm. How do we find *m* now? Well, just like we did in the previous section, if we divide the bigger number by the smaller one, we can find out what should go after the multiplication symbol. So what's 3 ÷ 2?

That's not hard: 3 ÷ 2 = 1.5, so now we know that *m* = 1.5 liters. (See p. 125 for a refresher on decimal division.)

Now we should double-check our answer to see if the cross products are equal:

$$\overset{3}{\underset{1.2}{2}} \overset{?}{\diagdown} \overset{3}{\underset{1.5}{2.5}}$$

Yep! The cross products are equal, so we found the right *m*. (And it makes sense that if you run a little bit more than 2 miles, you should need a little bit more than 1.2 liters of water, right?) Answer: 1.5 liters.

QUICK NOTE! In *proportions*, it's important to note that units must "mirror" each other in the two fractions. For instance, in the previous example, if we had been told that for every 2 miles, I drink 1.2 liters, but were then asked to find out how much water should I drink if I run 2.5 kilometers, we'd need to change miles to kilometers (km) or vice versa, so they'd be the same.

Step-By-Step

Using proportions in word problems:

Step 1. Create a "This *is* to that, *as* this *is* to that" sentence to help determine the proportion's fractions. Then, begin by setting up the proportion in words: $\frac{this}{that} - \frac{this}{that}$ (Of course, when working on a specific problem, you'll fill in the specific words from the problem.)

Step 2. Fill in the numbers you know, and plug in *m* for the missing number—be sure to include *m*'s units.

Step 3. Make sure the units "mirror" each other in the two fractions.

Step 4. Find the cross product, and set them equal to each other, still using *m* for the missing number.

Step 5. Multiply out the side of the equation that has two numbers, and then use division to find *m*.

Step 6. Plug *m* into the proportion, then check the cross products to make sure they are equal. Done!

Say we want to make some cookies for our friends. There's a peanut butter cookie recipe we found online that uses 2 cups of flour and makes 3 dozen cookies. But what if we only have $1\frac{1}{2}$ cups of flour? Assuming we have enough of all the other ingredients, how many cookies can we make?

Steps 1 and **2.** Set up the proportion in words: "2 cups of flour *is* to 3 dozen cookies, *as* $1\frac{1}{2}$ cups of flour *is* to how many cookies? We're trying to see how many cookies we can make with a certain amount of flour, and we'll use m for the missing number:

$$\frac{\text{cups of flour}}{\text{cookies}} = \frac{\text{cups of flour}}{\text{cookies}} \rightarrow \frac{2 \text{ cups}}{3 \text{ dozen cookies}} = \frac{1\frac{1}{2} \text{ cups}}{m \text{ cookies}}$$

Step 3. Make sure the units mirror each other. It might seem like they are the same, but actually, they're not—look at the bottom. On the left, we have "3 dozen cookies," and on the right we have just "cookies." We need to pick one and have both sides be the same.

Let's write out the 3 dozen cookies as 36 cookies.*

$$\frac{2 \text{ cups}}{36 \text{ cookies}} = \frac{1\frac{1}{2} \text{ cups}}{m \text{ cookies}}$$

Step 4. We're ready to cross multiply and set the cross products equal to each other:

$$2 \times m \qquad 36 \times 1\frac{1}{2}$$

$$\frac{2}{36} \diagtimes \frac{1\frac{1}{2}}{m}$$

It's our job to find the m that makes this statement true: $2 \times m = 36 \times 1\frac{1}{2}$

Step 5. Multiply out $36 \times 1\frac{1}{2}$. We recall from p. 53 that in order to multiply with a mixed number, we need to first convert it into an improper fraction. Using the MAD Face method, we get that $1\frac{1}{2} = \frac{3}{2}$. So, $36 \times 1\frac{1}{2}$ is the same as $36 \times \frac{3}{2}$. Now multiply: $\frac{36}{1} \times \frac{3}{2} = \frac{108}{2}$ (and now we reduce) $\frac{108 \div 2}{2 \div 2} = 54$

....

* *You know that 1 dozen equals 12 cookies, yes? That's why 3 dozen equals 36 cookies.*

But don't stop yet! This is just the cross product. We've just simplified $36 \times 1\frac{1}{2}$ to 54. Now we need to find out what m needs to be in order to make the cross products equal to each other.

$$2 \times m = 54$$

Hmm. So m, when multiplied by 2, gives us 54. Well, since $54 \div 2 = 27$, we know that $2 \times 27 = 54$. Aha! So m must be 27. Let's check the cross products to make sure we got it right.

$$\frac{2}{36} \overset{?}{\underset{\times}{}} \frac{1\frac{1}{2}}{27}$$

Yep! Both cross products = 54, so we found the right value for m. So $m = 27$. We now know that if 2 cups of flour will make 36 cookies, then $1\frac{1}{2}$ cups of flour will make 27 cookies. We can make a lot of cookies!

Watch Out! When you are figuring out what m should be, be careful, because you don't always divide the bigger number by the smaller one; it's always the number that's *by itself* on one side of the equation that gets divided *into*—even if it's a smaller number! For example, you might know that if you are trying to find m in the equation $4 \times m = 12$; you just divide $12 \div 4 = 3$, and $m = 3$.

But check this out: $4 \times m = 0.12$.

It may not be as obvious this time how to find m, but you still need to divide the number that's *by itself* by the 4:

$$0.12 \div 4 = 0.03$$

And $m = 0.03$. (I used decimal division here.) Also, always test your m afterward in the original equation, and you'll catch any mistakes if you make them!

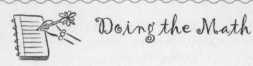 Doing the Math

Solve these word problems using proportions. I'll do the first one for you!

1. If 4 lipsticks cost $18, then how much would 5 lipsticks cost?

<u>Working out the solution</u>: We'd first say to ourselves: "4 lipsticks are to $18, as 5 lipsticks are to what?" and then set up the proportion in words:

$$\frac{\text{lipsticks}}{\text{dollars}} = \frac{\text{lipsticks}}{\text{dollars}}$$

Now, filling in the numbers: $\frac{4 \text{ lipsticks}}{\$18} = \frac{5 \text{ lipsticks}}{m \text{ dollars}}$

$$4 \times m \quad 90$$

$$\frac{4}{18} \diagdown \frac{5}{m}$$

Set the cross products equal to each other:
$4 \times m = 90$. What is m? Well, $90 \div 4 = 22.5$. So $m = \$22.50$. Double-check the cross product of our filled-in proportion:

$$90 \qquad 90$$

$$\frac{4}{18} \overset{?}{\diagdown} \frac{5}{22.50}$$

Yep, since the cross products are equal (both equal to 90), we've found the right m!

<u>Answer</u>: 5 lipsticks would cost $22.50.

2. Before your mom will let you go to your friend's house, you must first finish your reading assignment. You've read 36 pages in the last 30 minutes. You have 15 pages left. How long will it take you to finish?

3. You want to make a recipe that usually asks for 6 cups of flour, and $\frac{3}{4}$ teaspoon of salt. If you only have 2 cups of flour, how many teaspoons of salt should you use in the smaller version of the recipe? (Be sure to give your answer as a fraction, since that's how teaspoons are measured.)

4. You're watching your friend's puppies for the week while she goes on vacation with her family. You have 3 puppies, and they usually go through $\frac{3}{4}$ of a bag of puppy food a week. Now, you're watching 5 puppies. How many bags of puppy food will they eat for the week?

(Answers on p. 293)

Danica's Diary

ADVENTURES IN ONLINE SHOPPING

Last Christmas, I wanted to buy my friend Lori one of those pretty silver mesh rings from the Tiffany & Co. website. But I couldn't really tell from the site how "wide" the mesh was. My friend has very delicate fingers, and I wanted to make sure the ring wouldn't look too chunky on her finger.

I called the site's customer service number from the "contact us" link and spoke with a very nice service representative, who informed me that the ring was .344 inches wide. I thanked her for her help, got off the phone, and proceeded to grab my ruler so I could compare that width to the size of my finger. But then I realized: a ruler only shows increments of 1/16ths!

Through a little decimal-fraction manipulation and proportions, I was able to determine that .344 inches = $\frac{344}{1000}$ inches = approximately $\frac{5.5}{16}$ inches. So I counted off about $5\frac{1}{2}$ of those little 1/16th marks on the ruler, put it to my finger, and saw how wide it would be! I decided that the ring would suit her nicely.

I ordered the ring that day, and gave it to her a few weeks later. She loved it!

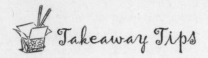 Takeaway Tips

- In proportions, the *cross products* will *always* be equal.

- To find the *missing number* in a proportion, you can "set" the cross products equal to each other, then find the missing number by figuring out what number will make the equality true. It's kind of like playing "what if": *what if* these two fractions made a proportion? *Then* what would *m* have to be, to make it true?

- When working with proportions, always make sure that your units *mirror* each other in the two fractions. If not, just do a unit conversion to convert whatever needs to be converted, and proceed as usual. (We'll discuss unit conversions in more detail in the next chapter.)

Chapter 19

Are You Drinking Enough Water?

Unit Conversions

Water is, hands down, the best thing you can drink. As I mentioned in the last chapter, it's great for your skin, but it's also great for your energy, your immune system—you name it! On the flip side, I'm a strong believer in staying away from sodas. Especially diet sodas. I've heard that diet sodas make you crave fatty foods, in fact, "There are several reports that the use of artificial sweeteners leads to an increased consumption of fat."* Yikes!

Anyway, they say that for a healthy body we should drink the same number of ounces in water every day as half our body weight (in pounds). So if someone weighs 110 pounds, they should drink 55 oz. of water a day. (Of course, always consult with your doctor before making radical changes to your diet.)

Personally, I prefer European bottled water, and they usually label

....

* "Can Artificial Sweeteners Help Control Body Weight and Prevent Obesity?" Benton, David. Nutrition Research Reviews, Volume 18, Number 1, June 2005, pp. 63–76(14). (Thanks to nutritionist Nancy Patourel for the reference!)

their bottles in terms of liters or centiliters. So how many *liters* is 55 oz.?

Well, if you know how many ounces make up one liter, that's a good place to start. There are a ton of websites (just Google "measurement conversion") that will tell you 1 liter ≈ 33.8 oz.*

But then what? How do we figure out how much 55 oz. is?

The easiest method I've found to convert one unit to another is to use handy little fractions called **unit multipliers**. Unit multipliers make this unit conversion stuff really easy, and you can use them in all sorts of circumstances, whenever there are different units involved.

In this case, we'd take the *unit multiplier* $\frac{1 \text{ liter}}{33.8 \text{ oz.}}$ and multiply it by the thing we're trying to convert: 55 oz.

Note that our unit multiplier fraction has two different *numbers* on the top and bottom, and also two different *units*—but the *amounts of water* described on the top and bottom are the *same*, because after all, 1 liter ≈ 33.8 oz.

Does this sound familiar: *when the top and bottom of a fraction are equal*? That's right: when the top and bottom are equal, it means that the fraction itself equals 1! And we know that we can always multiply a number by 1 and not change its value.

So, if we multiply 55 oz. by our unit multiplier, <u>we will change the units *without* changing the amount of water.</u>[†]

$$55 \text{ oz.} \times \frac{1 \text{ liter}}{33.8 \text{ oz.}}$$

As you know, before we can do fraction multiplication, we need to convert our whole number (55 oz.) into a fraction by putting it over 1, and then we multiply across the top and bottom, just like regular fraction multiplication:

$$= \frac{55 \text{ oz.}}{1} \times \frac{1 \text{ liter}}{33.8 \text{ oz.}} = \frac{55 \text{ oz.} \times 1 \text{ liter}}{1 \times 33.8 \text{ oz.}}$$

And now we actually get to *cancel* the *units* that are on both the top and bottom:

$$\frac{55 \text{ oz} \times 1 \text{ liter}}{1 \times 33.8 \text{ oz}} = \frac{55 \cancel{\text{ oz}} \times 1 \text{ liter}}{1 \times 33.8 \cancel{\text{ oz}}} = \frac{55 \times 1 \text{ liter}}{1 \times 33.8} = \frac{55}{33.8} \text{ liters}$$

....

* *The symbol ≈ means "approximately equal to." Your textbook might have a different approximation for this unit conversion.*

[†] *I'm going to do this problem one tiny step at a time, so you can see how it works. But in practice, these problems go very quickly—you'll see!*

Recall that fractions are division problems in disguise, so we can now just divide the top by the bottom.

$$= 55 \div 33.8 \text{ liters} \approx 1.63 \text{ liters}$$

Wow! That's more than $1\frac{1}{2}$ liters of water, every day!

To review: If a fraction has two different units and different numbers on top and bottom, but the total *amount* on top is equal to the total *amount* on the bottom, then you have a *unit multiplier*. The "units" in unit multipliers are just as important as the numbers!

What's It Called?

Unit Multipliers

Unit multipliers are fractions that equal 1 and that use two different units (and numbers) on the top and bottom of the fraction. For example: $\frac{12 \text{ inches}}{1 \text{ foot}}$ and $\frac{1 \text{ foot}}{12 \text{ inches}}$. (Note that fractions like $\frac{12 \text{ inches}}{1 \text{ foot}}$ and $\frac{60 \text{ seconds}}{1 \text{ minute}}$ would *not* be equal to 1 if we *didn't include their units*, since $\frac{12}{1} \neq 1$ and $\frac{60}{1} \neq 1$. For this reason, when dealing with *unit multipliers*, make sure to always include the units!)

QUICK NOTE! You may be wondering, "When do I use $\frac{12 \text{ inches}}{1 \text{ foot}}$, and when do I use $\frac{1 \text{ foot}}{12 \text{ inches}}$?" Good question. It all depends on what you're starting off with, and what you're trying to convert to.

Let's say you want to convert 432 inches into feet. Now, this is a simple one; you might just choose to divide it by 12. But if you use a unit multiplier, you would do $\frac{432 \text{ inches}}{1} \times \frac{1 \text{ foot}}{12 \text{ inches}}$ because you want the inches to <u>cancel out</u>, and you want the answer to be in feet.

You also probably now see why it's important to write the 432 inches as a fraction with a 1 on the bottom—so you can be sure of what's supposed to cancel with what.

If you had tried to do $\frac{432 \text{ inches}}{1} \times \frac{12 \text{ inches}}{1 \text{ foot}}$, for instance, then, as you can see, the inches wouldn't cancel! Oops.

Converting units using unit multipliers:

Step 1. Write down the *equality* that you'll use to make your unit multiplier. (For example: 12 inches = 1 foot.)

Step 2. Write the amount (with its unit) that you are trying to *convert from* as a fraction, with 1 on the bottom. (Like our $\frac{432 \text{ inches}}{1}$ from above.)

Step 3. Build a unit multiplier that has the same unit in the *bottom* of the fraction as the amount you're trying to *convert from*.

Step 4. Line up the two fractions, multiply, cancel your units, and divide to find your final answer!

And...
Action! Step-By-Step in Action

Your best friend just completed a 5K run (that's 5 kilometers) to help raise money for breast cancer research. How far did she run in miles?

Step 1. First, let's write down the *equality* that we'll use to build our unit multiplier: 1 mile ≈ 1.61 kilometers. (You can find this equality online, or refer to the chart on p. 232.)

Step 2. Write 5 km as a fraction, keeping its unit: $\frac{5 \text{ km}}{1}$

Step 3. Since we are trying to convert kilometers to miles, we'll want the *kilometers* to cancel out. So our *unit multiplier* should have *km* on the *bottom*: $\frac{1 \text{ mile}}{1.61 \text{ km}}$

Step 4. Line our fraction up with the unit multiplier, and cancel away! (Be sure that the unit you want to cancel is on top and bottom, just like canceling factors.)

$$\frac{5 \text{ km}}{1} \times \frac{1 \text{ mile}}{1.61 \text{ km}} = \frac{5 \text{ km} \times 1 \text{ mile}}{1 \times 1.61 \text{ km}} = \frac{5 \times 1 \text{ mile}}{1 \times 1.61} = \frac{5 \text{ miles}}{1.61}$$

Remember, fractions *are* division, so when you see something like this, just divide the top by the bottom and you can't go wrong! So $5 \div 1.61 \approx 3.11$.

Answer: She ran approximately 3.11 miles

QUICK NOTE! Remember, multiplying by 1 doesn't change the value of a number—this is why you can multiply numbers by *unit multipliers* and not change their value. All unit multipliers = 1.

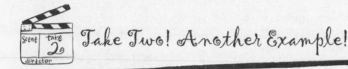

Take Two! Another Example!

There are 8 ounces in a cup, and 2 cups = 470 ml. How many milliliters are in 10 ounces?

Hmm, this one's a little tricky. It looks like we'll need to go from ounces to cups to milliliters. For this problem, we'll need *two* unit multipliers!

Step 1. Begin by listing the equalities that will make up the *unit multipliers*: 8 oz. = 1 cup, 2 cups = 470 ml.

Step 2. Next, write 10 oz. as a fraction: $\frac{10}{1}$ This is what we're converting *from*.

Step 3. We know that we want the oz. to cancel (since we are trying to convert 10 oz. to milliliters), so our *unit multiplier* must have oz. on the bottom: $\frac{1 \, cup}{8 \, oz.}$

And because we want to end up with milliliters, we'll want the cups to cancel, too. So in the next unit multiplier, let's put the cups on the bottom: $\frac{470 \, ml}{2 \, cups}$

Step 4. Now let's line everything up. . . .

$$\frac{10 \, oz.}{1} \times \frac{1 \, cup}{8 \, oz.} \times \frac{470 \, ml}{2 \, cups} = ?$$

. . . and cancel the units. (Don't you just love canceling?)

$$= \frac{10 \, oz.}{1} \times \frac{1 \, cup}{8 \, oz.} \times \frac{470 \, ml}{2 \, cups} = \frac{10 \, oz \times 1 \, cup \times 470 \, ml}{1 \times 8 \, oz \times 2 \, cups} = \frac{10 \times 1 \times 470 \, ml}{1 \times 8 \times 2} = \frac{4700 \, ml}{16}$$

Just divide the top by the bottom (to get the decimal answer): 293.75 ml. Voilà!

Note: You also could have done this in two steps—first converting from ounces to cups, getting an answer, and then converting from cups to milliliters for your final answer—but I wanted to show you how to use more than one unit multiplier like this. Remember, multiplying by unit multipliers just converts the units—not the amounts—so you can multiply by as many as you want and not change the value—just keep track of the units you are canceling to make sure you're doing it correctly!

Useful Unit Multipliers*

As I mentioned earlier, there are a ton of websites that have "unit equivalence" tables. For your convenience, though, here's a short list of some of the most common ones you'll come across. *Remember, the value of every one of these fractions = 1.*

Since 12 inches = 1 foot: $\frac{12 \text{ inches}}{1 \text{ foot}}$ or $\frac{1 \text{ foot}}{12 \text{ inches}}$

Since 3 feet = 1 yard: $\frac{3 \text{ feet}}{1 \text{ yard}}$ or $\frac{1 \text{ yard}}{3 \text{ feet}}$

Since 1 meter = 100 centimeters: $\frac{1 \text{ m}}{100 \text{ cm}}$ or $\frac{100 \text{ cm}}{1 \text{ m}}$

Since 1 inch = 2.54 centimeters: $\frac{1 \text{ inch}}{2.54 \text{ cm}}$ or $\frac{2.54 \text{ cm}}{1 \text{ inch}}$

Since 1 mile ≈ 1.61 kilometers: $\frac{1 \text{ mile}}{1.61 \text{ km}}$ or $\frac{1.61 \text{ km}}{1 \text{ mile}}$

Since 100 cents = 1 dollar: $\frac{1 \text{ dollar}}{100 \text{ cents}}$ or $\frac{100 \text{ cents}}{1 \text{ dollar}}$

Since 60 seconds = 1 minute: $\frac{60 \text{ seconds}}{1 \text{ minute}}$ or $\frac{1 \text{ minute}}{60 \text{ seconds}}$

Since 60 minutes = 1 hour: $\frac{60 \text{ minutes}}{1 \text{ hour}}$ or $\frac{1 \text{ hour}}{60 \text{ minutes}}$

Since 4 quarts = 1 gallon: $\frac{4 \text{ quarts}}{1 \text{ gallon}}$ or $\frac{1 \text{ gallon}}{4 \text{ quarts}}$

Since 2 pints = 1 quart: $\frac{2 \text{ pints}}{1 \text{ quart}}$ or $\frac{1 \text{ quart}}{2 \text{ pints}}$

Since 16 ounces = 1 pound: $\frac{16 \text{ oz.}}{1 \text{ lb.}}$ or $\frac{1 \text{ lb.}}{16 \text{ oz.}}$

Since 1 foot ≈ 0.305 meters: $\frac{1 \text{ ft.}}{0.305 \text{ m}}$ or $\frac{0.305 \text{ m}}{1 \text{ ft.}}$

· · · · · · · · · · · · · · · · · · · ·

* *Some of these unit multipliers are based on approximate equalities (whenever you see ≈), so those unit multipliers are only approximately equal to 1. But we still treat them as if they are equal to 1 because they're so close. Just be sure to use the ≈ symbol in your answer for those problems, to acknowledge this discrepancy!*

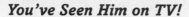

You've Seen Him on TV!

"*I love that you can actually have a conversation with a smart girl that doesn't involve your favorite clothing brand.*" Justin Chon, Tony Park on Nickelodeon's *Just Jordan* and Peter Wu on the Disney Channel's *Wendy Wu: Homecoming Warrior*

Creating Your Own Unit Multipliers

The unit multipliers above will certainly come in handy, but you can make your own unit multipliers, too! As long as the amount on top equals the amount on the bottom, the fraction will equal 1, and you can safely multiply with it and not change the value of the number whose units you're converting.

For example: Let's say that on Planet Glam, there is no money, just cool stuff. And you learn that 3 mascaras are worth the same as 4 lip glosses. You own a mascara company, and one of your orders produced a surplus of 126 mascaras that you don't need. Now you want to trade in all 126 mascaras. How many lip glosses can you get for them?

There are several ways to do this, but let's use *unit multipliers* to be sure we don't make a mistake.

What's our equality? 3 mascaras = 4 lip glosses.

So our *unit multiplier* could be $\frac{3 \text{ mascaras}}{4 \text{ lip glosses}}$ or $\frac{4 \text{ lip glosses}}{3 \text{ mascaras}}$

Both unit multipliers have equal tops and bottoms (which makes them both equal to 1), so they'd both work. The question is, which one will be more useful to us? Well, we want to multiply 126 mascaras times something that will cause the units to cancel, so we should use a unit multiplier that has *mascaras* on the bottom, and remember from fraction multiplication (see p. 51) that we always put whole numbers over 1 before multiplying:

$$126 \text{ mascaras} \times \frac{4 \text{ lip glosses}}{3 \text{ mascaras}} = ? \rightarrow \frac{126 \text{ mascaras}}{1} \times \frac{4 \text{ lip glosses}}{3 \text{ mascaras}} = ?$$

Now we can *cancel* the "mascara" units, and just multiply like normal fraction multiplication across the top and bottom:

$$\frac{126 \text{ mascaras}}{1} \times \frac{4 \text{ lip glosses}}{3 \text{ mascaras}} = \frac{126 \cancel{\text{ mascaras}} \times 4 \text{ lip glosses}}{1 \times 3 \cancel{\text{ mascaras}}} = \frac{126 \times 4 \text{ lip glosses}}{1 \times 3}$$

Before we do the multiplication, however, we may notice that $1 + 2 + 6 = 9$, so 126 is divisible by 3. So let's first reduce by canceling a "3" from 126 and 3. (See p. 9 for divisibility tricks!)

$$\frac{126 \times 4 \; lip \; glosses}{1 \times 3} = \frac{\overset{42}{\cancel{126}} \times 4 \; lip \; glosses}{1 \times \cancel{3}_{\,1}} = \frac{168 \; lip \; glosses}{1}$$

So, the answer is 168 lip glosses. Not bad!

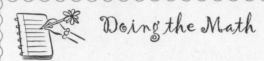 *Doing the Math*

Convert these units, using either *unit multipliers* from the chart on p. 232, or by building your own unit multiplier based on the information given. I'll do the first one for you.

1. How many centimeters is 4 feet?

<u>Working out the solution</u>: We know from the chart that 2.54 cm = 1 inch, and that 12 inches = 1 foot, so let's use two unit multipliers to convert from feet to centimeters. Since we want the feet to cancel out, we should make a unit multiplier with feet on the *bottom*: $\frac{12 \, in.}{1 \, ft.}$ Then, since we'll want to end up with centimeters in our answer, we'll want the second unit multiplier to have cm on top: $\frac{2.54 \, cm}{1 \, in.}$ Now let's write 4 ft. as a fraction over 1, $\frac{4 \, ft.}{1}$, put it all together and cancel away!

$$\frac{4 \, ft.}{1} \times \frac{12 \, in.}{1 \, ft.} \times \frac{2.54 \, cm}{1 \, in.} = \frac{4 \, \cancel{ft.} \times 12 \, \cancel{in.} \times 2.54 \, cm}{1 \times 1 \, \cancel{ft.} \times 1 \, \cancel{in.}} = \frac{4 \times 12 \times 2.54 \, cm}{1} = 121.92 \, cm$$

<u>Answer</u>: 4 ft. = 121.92 cm

2. How many kilometers is 6 miles?

3. How many meters is 5 feet?

4. If 3 purses can be traded in for 10 bottles of nail polish, then how many bottles of nail polish can you trade in for 42 purses?

5. How many gallons is 36 pints? (Hint: Use two unit multipliers!)

(Answers on p. 293)

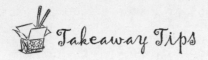 *Takeaway Tips*

- Unit multipliers *all have a value of 1.*

- Unit multipliers *take you from one unit to another;* they don't change the *amount* of the thing you're converting, just the *units* that the amount is expressed in.

- Always construct your unit multipliers so that the units cancel in the way you want them to. Usually, *the unit you want to cancel out* (the unit you are converting *from*) *should be on the bottom* of the unit multiplier.

- When using unit multipliers, units *cancel away just like factors* do.

Quiz #3: What's Your Learning Style?

Did you know that everyone learns differently? Let's see what expert psychologist Robyn Landow, PhD, says about *your* learning style. Take her quiz and see how you fare!

1. You're new at school, and you need directions to the cafeteria. You approach a friendly-looking student who you chatted with that morning. You ask her:

    a. To tell you the directions.

    b. To draw a quick map on a piece of paper.

    c. To walk you there.

2. You like websites that:

    a. Have audio channels where you can hear music or interviews.

    b. Have an interesting design and graphics.

    c. Keep you busy. You like to click around, participate in polls, etc.

3. Your teacher asks you to give the class a short talk about your rock-climbing hobby. You:

    a. Tell some great stories and make the class feel like they are on that mountain with you.

    b. Bring in some great footage of your most recent climb, so they can see what it was like for themselves.

    c. Bring in your gear and help your classmates try it on to get a feel for what it's like to climb.

4. Your parents have finally agreed to buy you a cell phone! While your mom is concerned about price, what matters the most to you?

    a. The salesperson's description of the phone's features.

    b. The design / if it looks cool.

    c. Trying it out and seeing how easy it is to use.

5. When your mind wanders during an oh-so-boring school assembly, you:

    a. Sing that hot new song in your head. What is that third line all about anyway?

    b. Read. It doesn't matter what it is, it has to be more interesting than this.

    c. Doodle in any available notebook, even if it's not yours!

6. You're about to get a puppy—finally! But your mom has made it clear: the puppy will be *your* responsibility. You've never been in charge of a dog before, so you:

    a. Speak to the experts at the local pet store. They can tell you everything you need to know.

    b. Run out to the bookstore to buy *The Ultimate Guide to Caring for your Puppy.*

    c. Go to your best friend's house. She's got a dog, and she'll let you help her while she feeds and walks him. Maybe you'll even help give him a bath!

7. You just got this great new digital camera for your birthday, and you want to learn how to use it ASAP, so you can take pictures of you and your friends at Friday night's bash. The first thing you do is:

    a. Call your best friend. She has the same camera, and you want to hear from her how to use it.

    b. Go to the camera's website. You heard the instructions are great, and they show examples of good and poor photos and how to improve them.

    c. Just start taking pictures. You're not afraid to get some bloopers before you get the hang of it.

8. When studying alone, you:

    a. Find yourself talking out loud in order to memorize something.

    b. Read your notes over and over, and sometimes rewrite them, just to get them solid in your mind.

    c. Create mock questions and then answer them.

9. Admit it, you are famous for being:

    a. A great conversationalist. So easy to talk to!

    b. An avid reader. You are the one to come to for a book recommendation!

    c. An artist or athlete. Whether it's dance, sculpture, or soccer, you are always at class or practice, putting your talent to use!

10. When trying to remember a phone number, you close your eyes and:

    a. Recite it in your head, as if you were speaking the numbers so you can "hear" it.

    b. Visualize the numbers on a piece of paper so you can "see" it.

    c. Let your "fingers do the walking" by visualizing yourself dialing the number.

11. It's your little sister's birthday next week, and you want to surprise her with a gift "from the heart." You are most likely to:

a. Write her a silly song to the tune of the latest number one hit and sing it at her birthday dinner.

b. Write a poem and put it on her nightstand so it is there when she wakes up in the morning.

c. Take her to see the latest chick flick—bonding time for just the two of you.

12. You signed up to help organize the school dance and are asked what committee you'd like to be on. You choose:

a. The music committee. You want to make sure the joint is jumping that night!

b. The decorations committee. The place has to look perfect, and you've got some great ideas!

c. The logistics committee. You want to make sure the night runs smoothly and be the "go-to" person should any glitches occur. After all, you're a great "people person."

Count up the number of *a*'s, *b*'s, and *c*'s you scored.

If you scored mostly *a*'s, you are an **auditory learner**! You learn best when information is presented in an auditory format; that is, you're a great listener! In the classroom, you benefit from listening to lectures and participating in group discussions. When trying to remember something, you can often "hear" the way someone told you the information or the way you previously repeated it out loud.

Learning strategies include:

- Always read your math problems out loud when you're doing your homework, and then *talk out* the solution. Like, "Let's see, if I'm supposed to divide this by that, then I should find the reciprocal . . ." Talking this stuff out will help you to stay focused.

- Create auditory review "notes" by reading and recording notes and textbook information out loud while you're studying—just the most useful stuff that helps you remember how to do certain problems. When preparing for a test, listen to the recordings.

- Create silly dialogues, or even turn them into songs, to help you remember math concepts. For example, whenever you see a mixed number you have to convert into an improper fraction, you can think, *"Mixed number? That makes me MAD, so I guess I'll use the MAD Face method."*

If you scored mostly *b*'s, you are a **visual learner**! You learn best when information is presented visually, either in a written language or picture/design format. In the classroom, you benefit from instructors who use the blackboard or "smartboard" to demonstrate the essential points of a lecture, or who provide you with an outline to follow along with during class. You tend to like to study by yourself in a quiet room. You often "see" information in your mind when you are trying to remember something. You may have an artistic side that enjoys projects having to do with visual art and design.

Learning strategies include:
- "Color code" your notes or textbook when studying. Using highlighter pens, highlight different kinds of information in contrasting colors.

- Write out sentences and phrases that summarize key information obtained from your textbook and lecture.

- Make flashcards of equations that need to be memorized. Limit the amount of information on the card so your mind can take a "mental picture" of the information.

- When a problem involves a sequence of steps, write out in detail how to do each step.

- Rewrite your notes from each day's class. Just *rewriting them* and *seeing* the information again will help to cement it in your head.

- Before a test, make visual reminders of information that must be memorized. Make "stick it" notes containing key words and concepts and place them in highly visible places—on your locker, your mirror, or your planner.

If you scored mostly *c*'s, you are a **tactile/kinesthetic learner**! You learn best when physically engaged in a hands-on activity. In the classroom, you enjoy lab settings where you can manipulate materials to learn new concepts. You prefer teachers who encourage in-class practice and demonstrations as well as "fieldwork" outside the classroom.

Learning strategies include:

- Think up ways to make your learning "real." Whenever possible, work out problems with actual items from around the house or make a model that illustrates a key concept. For instance, in the equivalent fractions chapter (chapter 6), you could actually order a pizza and try cutting it in different ways to express equivalent fractions. Actually doing the cutting *yourself* will help to make it real.

- To stay focused in class, sit near the front of the room and take notes to stay involved.

- When studying, try walking back and forth with notes/flashcards in hand and reading the information out loud. Then recite them while taking a shower, jogging, or doing other physical activities. This will help the information "get in your body."

- When memorizing a sequence for how to solve a problem (like the "Step-by-Step" lists throughout this book), make flashcards for *each individual step*. Arrange the cards on a table to represent the correct sequence. Put words, stickers, symbols, or pictures on the flashcards to "personalize" them. Practice putting the cards in order until you can do it with ease.

- When appropriate, *act out* the concepts or weave them into stories, so you can really get a feel for it. For example, when memorizing the "division" definitions (see p. 126), you could pretend to *be* the person standing on the rooftop, saying, "Sparky! You'll get dinner when it's ready" and, "I'm the quotient, because I'm getting quoted." And then pretend to be Sparky, *eyeing* the steak inside, and really act out the "eyeing," saying, "Now I'm the div-EYE-sor." Make up your own stories to act out, too—use your creativity to make studying more interesting and effective!

Chapter 20

Who's the Cute New Foreign Exchange Student?

Introduction to "Solving for x"

$\mathcal{B}$efore we get into solving for x—and cute foreign exchange students—check out the multiplication symbols box below. Different books have different preferences regarding which multiplication symbol to use, and I want to make sure you're never confused by all the different symbols out there!

Multiplication Symbols

For most of this book, I've used the symbol $\times$ to represent multiplication, but there are a few other symbols which you might see as well: $\cdot$, ( ), and sometimes there is no symbol at all! (This is the most common type in algebra.)

For example, you can write "3 times m" in a few different ways:

$$3 \times m$$
$$3 \bullet m$$
$$3(m)$$
$$3m$$

Yes, these are all different ways of saying the same thing, and you should be familiar with them, because you never know which one you'll see. (Kind of annoying, I know, but don't worry—at a certain point you'll get used to switching back and forth between them.)

Placeholders and Nicknames

Let's say a new student comes into your classroom—a really cute guy from a foreign country. He sits down next to you, and tells you his name is Vakhtangi Levani Gachechiladze. *Ohmigod* he's so cute. Wait—*what* was his name? Panic sets in. "How will I ever remember it? How will I introduce him to my friends? What if I forget it at the wrong time?"

Then he says, "But you can call me V."

And you are breathing again. What a relief! Even if you could re-member it, you would surely have mispronounced it. Let's face it: if you can't pronounce his real name and you can't remember his real name—even though he's *super* cute—you don't *know* his real name.

Sometimes in math, when we don't *know* the real value of some-thing, we give it a "nickname." Like in chapter 18, when we were working with proportions, we called the missing number m. That was a placeholder or a nickname. But we could have also used V or, heck—any letter we wanted! **Algebra** uses nicknaming all the time.

Say you get one of those "free gift with purchase" makeup bags. There are a bunch of lipsticks included in the bag, but you don't know exactly how many. Now let's say you give your bag to me, and I mag-ically make the amount of lipsticks double, then I take 3 lipsticks out of the bag, and hand it back to you. (Just go with me on this.)

Let's describe how many lipsticks are in your bag *now*, compared to how many you *started with*, using a nickname. Let's call the amount you started with "S." ("S" stands for "Start," since it's the starting amount.)

When I magically made the number of lipsticks double, we would have 2 times S, right? That could be written $2 \times S$, or $2S$. Then I subtracted 3 lipsticks from it, so it would become "$2S - 3$."

See how neat and concise that is? And if you know how many lipsticks you started off with (i.e., the value of "S"), then "$2S - 3$" perfectly describes the number of lipsticks you have *now*.

This example demonstrates the important skill of using the language of math to write a "math expression" from a situation described in regular words. File this idea away in the back of your mind, and as you study more and more math, you'll see how this kind of thinking comes in handy.

Doing the Math

Write an expression using "nicknames." I'll do the first one for you.

1. I am holding a bunch of sticks of gum, but I'm not going to tell you how many. Let's call the number of sticks of gum "g." Let's say that you give me 4 more sticks, and then I give you half of everything I have. Written in terms of g, how many sticks of gum did I give you?

<u>Working out the solution</u>: Let's take this problem step-by-step. Okay, I start out with g pieces of gum. Then you give me 4 more pieces. Now I have $g + 4$ pieces. Then I give you half of everything I have—in other words, half of $g + 4$. So we could either write this out as $(g + 4) \div 2$ or as $\frac{g+4}{2}$.

<u>Answer</u>: $(g + 4) \div 2$ or $\frac{g+4}{2}$.

(Remember, fractions "are" division! See p. 42 for a review of that helpful factoid.)

2. I have a bag of beads, and there is a certain number of beads in it—but I'm not telling you how many. Let's just call the number of beads in the bag right now "x." What if I then add 3 beads to it? How many beads do I have now?

3. You gave your sister a box of mints; we'll call the number of mints in the box "m." She took 2 of them, and handed the box back

to you. Then you ate 5 more. In terms of *m*, how many mints are now left in the box?

4. For Christmas, your sister gave you a box of your favorite chocolates. Let's call the number of chocolates in that box "*c*." You ate 3 of them—and then your mom gave you another, identical box of chocolates! In terms of *c*, how many chocolates do you have now?

(Answers on p. 293)

In algebra, nicknames like *x*, *g*, *m*, etc., are called **variables**, because they can all stand for a *variety* of amounts. Just like *V* could stand for Victor, Vakhtangi, Vampires, or whatever, the variable *x* could stand for 2, 8, 1001, or any other conceivable value. The point is, we use these nicknames—these variables—as placeholders because we don't know their real value—yet.

What's It Called?

Variable

A *variable* is a letter that stands for a number whose value we don't know yet. It's a "nickname" for an unknown number. The most famous variables are *x* and *y* (you'll see them all the time in algebra), but you can use whatever letter(s) you'd like.

One of the main reasons mathematicians use variables in "math sentences" is because they're such an efficient way to "speak"!

Bag of Pearls

When you see *x* or some other *variable* in an expression, you can think of that variable as a bag of pearls.

There is *some* number of pearls in the bag—you may not know how many, but that doesn't mean the number doesn't exist. It's just a number *you* don't happen to know yet. That's why we call it *x*.

If *x* is 1 bag of pearls, then what is 2*x*? It's 2 bags of pearls.

It's important to remember that both bags are the same: they each have *x* pearls inside them!

What is 2*x* + 3? It's 2 *bags* of pearls, plus 3 loose pearls.

Even though you don't know how many pearls are in each bag, if I told you that this is true:

$$2x + 3 = 13$$

then with a little algebra, you could find out how many pearls are in each bag—without ever opening them. Let's write out that equation in pearls.

What we have here is a true statement: an equation. Both sides of the equation are equal to each other. In other words, there are the *same number of total pearls* on the left side as the right side: they weigh the same amount as each other, so we have a balanced scale. That's what = means. (Imagine that the bag material is very light and, for all practical purposes, doesn't weigh anything.)

Our *goal* is to have a scale with *one bag* on one side, and some number of loose pearls on the other. How do we do that? Let's see. If we take away 3 pearls from each side, the scale will still be balanced, right?

This is the same as subtracting **3** from both sides of the equation, see?

$$2x + 3 - \mathbf{3} = 13 - \mathbf{3}$$
$$\rightarrow$$
$$2x = 10$$

We now know that 2 bags of pearls are equal to 10 loose pearls. But we want to know how much is in 1 bag, not 2. So what if we divide both sides by 2? That's the same thing as saying we want to divide both sides in *half*. Half of two bags, 2*x*, would just be one bag: *x*. And half of the right side, 10, would be 5.

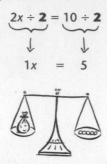

$$2x \div \mathbf{2} = 10 \div \mathbf{2}$$

$$1x \ = \ 5$$

One bag equals 5 pearls.

Amazing! We figured out how many pearls must be in the bag, without ever opening it.

And we've just done a real live algebra problem.

QUICK NOTE! 1*x* is the same thing as *x*. They both mean "one bag of pearls."

Boxes vs. Bags?

When I first learned this stuff, I used to like to draw in a little box for *x*, and then solve the equation. For some reason, this made it easier for me to understand what things like 4*x* meant.

$$X \ = \ \square$$

I knew that I could subtract 4□ − 1□, for example, and get 3□. Or that I could divide 4□ ÷ 4 = 1□. After all, what is 4 boxes minus one box? It's 3 boxes. And what is 4 boxes, divided into 4 parts? Well, 1 box! (I know that we talked about a *bag* of pearls above, but a *box* is easier to draw than a bag!)

So feel free to use a □ instead of *x* when solving these types of equations. And just imagine that there is some number of pearls inside that box—and it's your job to figure out how many!

Solving for X

> When a math problem asks you to find the value of *x*, the goal is to *isolate x*, by doing things to both sides of the equation—until *x* is *all by itself* on one side, and there is a *number* on the other side—and that number is your answer!

Step-By-Step ♀ ♂

Solving for x:

Step 1. Rewrite the equation with boxes instead of *x*'s, if it helps.

Step 2. Do "things" (add, subtract, multiply, divide) to both sides of the equation in order to get all the stuff with the *x*'s (or boxes) on *one* side, and all the plain *numbers* on the *other*. Remember, you must do the same "things" to both sides, to keep the scales balanced at all times, or the two sides won't equal each other anymore.

Step 3. Then do "things" to both sides of the equation in order to get just *one x* (or one box) by itself on one side, and just *one number* on the other side, so that the equation now looks like $x = $ number. Voilà! You've solved for *x*!

Dividing on Both Sides of the Equation . . . Using Fractions!

Usually, in prealgebra and algebra, when we divide both sides of an equation by a number, it's so that we can isolate the *x*. You can use the ÷ symbol, like we did in the previous example, but as the problems get more complicated, the easiest way to divide—believe it or not—is by using *fraction notation.** I'll show you how this works:

For the equation $2x = 10$, we'd want to divide both sides by 2, right? Rather than use the ÷ symbol, we could instead divide both sides by drawing *fraction lines* under them, making them the numerators in fractions, using 2 as their denominators. As long as we do the exact same thing to both sides of the equation, we'll keep our scales balanced!

....
* *Remember that "fractions are division"!*

$$\frac{2x}{2} = \frac{10}{2}$$

Now, 2 and x are just factors in a fraction. And we can reduce the fraction by canceling a 2 from the top and bottom.

$$\frac{\cancel{2}x}{\cancel{2}} = \frac{\cancel{10}^{5}}{\cancel{2}}$$
$$\downarrow \qquad \downarrow$$
$$x = 5$$

We've solved for x, and everything we did was perfectly allowed. That is, we kept the scales balanced at all times, so what we ended up with, $x = 5$, is indeed an equality.

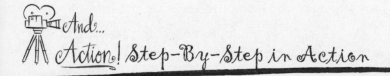

And... Action! Step-By-Step in Action

Solve for x. (In other words, find the x that would make this equality true.)

$$4x - 1 = 19$$

Step 1. First, we'll rewrite the x as a little box.

$$4\square - 1 = 19$$

Step 2. In order to get the $\square$ by itself, which is our goal, the first thing we should do is get rid of that pesky -1 so that at least $4\square$ will be by itself on one side of the equation—we'll deal with the $4\square$ later. How can we get rid of the -1? Let's *add* 1 to both sides.

$$4\square - 1 + \mathbf{1} = 19 + \mathbf{1}$$
$$4\square = 20$$

Step 3. This latest form of the equation says that 4 boxes of pearls *equal* 20 pearls total. You could probably guess that each box would have 5 pearls in it—but here's what we should do: we divide both sides of the equation by 4, to "get rid of" the 4, and then cancel common factors on top and bottom.

$$4\square = 20$$

$$\frac{4\square}{4} = \frac{20}{4}$$

$$\frac{4\square}{\cancel{4}} = \frac{\cancel{20}^{5}}{\cancel{4}}$$

$$\square = 5$$

Now remember that $\square$ is the same as x, so let's plug in $x = 5$ into the original equation to see if it works:

$$4x - 1 = 19$$

$$4 \times 5 - 1 \overset{?}{=} 19$$

$$\rightarrow \quad 20 - 1 \overset{?}{=} 19$$

$$\rightarrow \quad 19 = 19 \quad \checkmark$$

Yep, $x = 5$ is our solution!

QUICK NOTE! Whenever we're looking at an equation, and we're supposed to solve for x, what we're really doing is trying to answer the question: "What does x have to be in order for this equation to be a true statement?"

Watch Out! Make sure to keep your pearls and boxes of pearls *separate*. If you have $4x - 1$, some may be tempted to "simplify" it like this: $4x - 1 \rightarrow 3x$, but this would be *wrong*! There's no way to simplify $4x - 1$, unless someone gives you a numeric value for x to plug in.

On the other hand, if you have $4x - 1x$, then you can subtract them: $4x - 1x = 3x$. Because, 4 boxes − 1 box = 3 boxes. See the difference?

Solve for x.

$$\frac{x}{3} + 2 = 5$$

Step 1. First, we'll rewrite this with a box: $\frac{\square}{3} + 2 = 5$. And what does $\frac{\square}{3}$ mean? It means "one third of a box of pearls." What *we* want to find is the value of □ by itself—the whole box. Don't worry. We can do this. Let's take it one step at a time.

Step 2. Remember, we're trying to get the □ by itself, so let's start by subtracting 2 pearls from each side, so that "all the stuff with the boxes" is on one side, and all the plain numbers are on the other.

$$\frac{\square}{3} + 2 = 5$$
$$\frac{\square}{3} + 2 - \mathbf{2} = 5 - \mathbf{2}$$

(subtract 2 from both sides, keeping the scales balanced)

$$\frac{\square}{3} = 3$$

Step 3. It's looking better—we seem to be getting *closer* to getting the □ all by itself, but we're not there yet. How can we get rid of the 3 on the bottom? *Hmm.* What if we multiplied the box fraction by 3? As long as we do that to *both* sides of the equation, we'll keep the scales balanced:

$$\frac{\square}{3} \times \mathbf{3} = 3 \times \mathbf{3}$$

(multiplying both sides by 3 keeps the scales balanced). Now we get:

$$\frac{\square \times \cancel{3}}{\cancel{3}} = 9$$
$$\square = 9$$

In other words: x = 9.

(You may be wondering how we got so "lucky" that the 3s cancel to get x by itself, but remember, that's exactly *why* we chose to multiply by 3.)

And we can easily check our answer by plugging **9** in for **x**, in the original equation: $\frac{x}{3} + 2 = 5$

$$\frac{9}{3} + 2 \overset{?}{=} 5$$

$$\rightarrow \quad \frac{\cancel{9}^{3}}{\cancel{3}_{1}} + 2 \overset{?}{=} 5$$

$$\rightarrow \quad 3 + 2 \overset{?}{=} 5$$

$$\rightarrow \quad 5 = 5 \quad \checkmark$$

Yep! We found the value of x that makes the original equation true, $x = 9$, and that's what we were trying to do!

QUICK NOTE! By doing the same thing to both sides of the equation, we ensure that, at each step along the way, we have a true equality. That way, when we finally manage to get x by itself on one side, we will *still* have a true equality. That is, $x = $ number!

Take Three! Yet Another Example!

Solve for x.

$$2x + 3 = 3x + 1$$

Hmm, now x is on both sides.

Step 1. Let's rewrite it with boxes: $2\square + 3 = 3\square + 1$

Step 2. So, two boxes of pearls plus 3 loose pearls is the same as 3 boxes of pearls plus 1 loose pearl? Once again, our goal is to get <u>one</u> box *by itself* somehow, by doing things to both sides of the equation, to keep both sides equal at all times.

How do we isolate one box? What if we take away two boxes of pearls from both sides? In other words, what if we subtract 2□ from both sides?

$$2□ - \mathbf{2□} + 3 = 3□ - \mathbf{2□} + 1$$

$$3 = 1□ + 1$$

Step 3. We're getting closer. The stuff with boxes is all on one side. But we still have to get the numbers on the other side. So, let's subtract 1 from both sides:

$$3 - \mathbf{1} = 1□ + 1 - \mathbf{1}$$

$$2 = 1□$$

And since 1□ is the same as □, we can say □ = 2, in other words, $x = 2$.

Now we can try plugging in 2 for x into the original equation, and make sure we got the right answer:

$$2\mathbf{x} + 3 = 3\mathbf{x} + 1$$

$$\overset{?}{\bullet}$$

$$2 \times \mathbf{2} + 3 = 3 \times \mathbf{2} + 1$$

$$\to \qquad 4 + 3 \overset{?}{\bullet} 6 + 1$$

$$\to \qquad 7 = 7 \quad \checkmark$$

Yep! So, 2 is the value for x that makes the equality true! Answer: $x = 2$

Do whatever you have to do in order to get x (or the □) by itself on one side of the equation and a number on the other. Add things, subtract things, multiply, divide, and so on. As long as you do the same thing to *both sides of the equation*, you'll be fine!

Watch Out! When you're "doing things" to both sides of an equation in order to move all the stuff with x's to one side and all the plain numbers to the other, at some point, you may make the mistake of dividing or multiplying too soon. Since you can always do "anything" to both sides of the equation and still keep an equality, you can't do much damage—but you must do it correctly. Here's an example of how you might do things out of order:

$$2x + 1 = 5$$

The *best* first step to isolate *x* is to subtract 1 from both sides. But let's say, instead, you wanted to divide both sides by 2. That's okay, but it makes things a little more complicated—you'll just have to be extra careful that you're doing it correctly. Remember: when you do something to an equation, you have to do it to *the entire side* of the equation. Here goes:

$$\frac{2x + 1}{2} = \frac{5}{2}$$

That would be fine, but it's not the most efficient way to isolate *x*, because (as you can see) we've just made things more complicated. But what you really *don't* want to do here is divide *just* the 2x by 2 and ignore the 1.

$$2x + 1 = 5$$

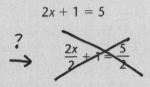

This is wrong!*

To avoid mistakes, your best bet is to *first* put everything with *x*'s on one side and all the plain numbers on the other, and then do multiplication or division to get just one *x* by itself.

"*I* used to be really good at math, but it's harder now. I want to get better at it." Elle, 11

"*I* like math when I have a fun teacher. My teacher has to be really supportive and sweet to me. I used to feel so insecure in math . . . I still sometimes feel stupid. I hate feeling stupid." Cindy, 15

What Do You Have to Say?

. . . .
* If you're curious to see how this problem would correctly be finished, starting out by dividing instead of by subtracting, see mathdoesntsuck.com.

Doing the Math

Find the x that makes these equalities true. (Feel free to change them into boxes first, if you'd like—and just think of the *variable* as being a bag or box of pearls!) I'll do the first one for you.

1. $5x = 3x + 8$

<u>Working out the solution</u>: First, we change the x's to boxes: $5\square = 3\square + 8$. Then we subtract 3 boxes from each side: $5\square - 3\square = 3\square - 3\square + 8$. Now we have $2\square = 8$. Divide both sides by 2, and we get:

$$\frac{\overset{}{\cancel{2}}\square}{\cancel{2}} = \frac{\overset{4}{\cancel{8}}}{\cancel{2}}$$

We are left with $\square = 4$. To check our answer, we can substitute 4 for x in the original equation and then $5x = 3x + 8$ becomes:

$$5 \times 4 \overset{?}{=} 3 \times 4 + 8$$

$$\rightarrow \quad 20 \overset{?}{=} 12 + 8$$

$$\rightarrow \quad 20 = 20 \ \checkmark \qquad \text{Answer: } x = 4$$

2. $x - 7 = 11$

3. $2x + 6 = 10$

4. $\frac{x}{5} + 1 = 3$

5. $8x = 7x + 5$

6. $6x + 1 = 2x + 5$

(Answers on p. 293)

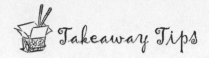 Takeaway Tips

- We use "nicknames" or variables, like *x* and *y*, because *we don't know what their value is yet*. They *have* a value, *we* just don't know what it is yet!

- Think of *x as a box or a bag of pearls*—in fact, you can plug in "boxes" instead of *x*'s, while working on the equation, if you prefer.

- The goal is to get *x by itself* by doing things to *both sides of the equation*—then the scales will remain balanced and you'll end up with a true statement at the end, in the form of: *x* = number.

- The best way to "solve for *x*" is to first get all the stuff with *variables* on one side (*x*'s, *y*'s, or boxes), and get all the *plain numbers* on the other. It's all about doing what you have to (to both sides of the equation) so that you get *x isolated* on one side of the equation—all by itself—then you've solved for *x*!

TESTIMONIAL:

Lisa Mays (West Chester, PA)

<u>Before</u>: Struggling, self-critical worrier
<u>After</u>: Successful neuroscience major at University of Southern California!

When I was in middle school, I remember feeling torn between wanting to do well and being afraid of seeming like a "nerd." In math class, students would always compare test grades to see who got the highest score, but at the same time most students would lie about the amount of studying they had done for the test because they didn't want to seem dorky. It was silly, because everyone knew deep down that the only way to succeed was to study hard.

Since I did study hard (whether or not I admitted it!) during my senior year of high school, I ended up being named a National Merit Finalist and getting accepted to a great college on a half-tuition scholarship. I'm so glad I did, because I love being a neuroscience major at USC!

Way Too Hard on Myself

What's weird about all of this is that even though I earned A's in math classes, I used to say (and think) that I wasn't very good at math. I still feel that way sometimes. I'm not completely sure why, but I think it has something to do with the fact that I get really mad at myself when I mess up simple parts of a complicated math problem—addition and subtraction, for example—which I often do. I used to always be really hard on myself with these kinds of mistakes (and sometimes I still am), and I know my self-confidence suffered for it. Something we all have to watch out for!

"Nobody's perfect—we can only try our best. When we do, we have a lot to feel good about!"

One thing I've learned is that you should never give up or feel inferior if you are struggling with something—everyone has his or her strengths and weaknesses. If you are having trouble with your math homework, find some friends who are having trouble as well, and work together, so you don't feel like you're the only one having problems.

Remember: Nobody's perfect—we can only try our best. When we do, we have a lot to feel good about!

Chapter 21

Romeo and Juliet

Introduction to "Solving for x" in Word Problems

$\mathcal{T}$here are some classic, timeless combos that have always inspired warm and fuzzy feelings: Romeo and Juliet, chocolate and peanut butter, mashed potatoes and gravy, solving for x and word problems . . .

Well, maybe not *so* many warm and fuzzies for that last one.

But guess what? It's not as bad as it sounds. And knowing how to solve for x can really be helpful when dealing with a complicated word problem. It can actually make the problem much *easier* to solve!

Remember this chart from page 192? We first saw it when we were learning how to "translate" English into math. We can now add *variables* like x and y to it, too. Throughout this chapter, this chart will come in handy for translating English into "math sentences" that include variables.

English	Math
of (only when surrounded immediately by two numbers)	× (multiplication)
per, quotient, a	÷ (division)*
sum, and, total, more than	+ (addition)
difference, less than	− (subtraction)
is, are	= (equals sign)
what, how much	x or y or m, etc. (a "nickname" for a number whose value we don't know yet)

Take a sentence like: "What is 3% of 30?" You can translate this whole thing, practically word for word, into math:

what → y (it's the thing we don't know)
is → =
of → × (since it's immediately surrounded by two numbers)

So we get:

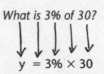

What is 3% of 30?

$$y = 3\% \times 30$$

And now, when you solve the problem, you're solving for y. To solve for y, first convert 3% to a decimal, and then solve $y = 0.03 \times 30 = 0.9$. Answer: $y = 0.9$

I know what you're thinking: "Why do I need to use x or y? Why can't I just solve it?" Well, you can . . . but practicing in this way will help to get you ready for algebra!

Here's another example:

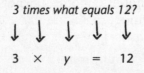

3 times what equals 12?

$$3 \quad \times \quad y \quad = \quad 12$$

....

* *When you can replace* a *with* per *in the sentence and it doesn't change the meaning, then* a *indicates division.*

Which is the same as $3y = 12$ (see multiplication symbol box on pp. 241–2). From what we've learned about solving for x, we can now solve this by rewriting the y as a box (because really, we can use any nickname we want!).

$$3\square = 12$$

Divide both sides by 3, and cancel the 3s:

$$\frac{\cancel{3}\square}{\cancel{3}} = \frac{\cancel{12}^{4}}{\cancel{3}}$$

So, $\square = 4$. Answer: $y = 4$

Free Gift with Purchase . . . Revisited!

Remember the example of the lipsticks in the "free gift with purchase" bag from the last chapter? Let's look at it in "word problem" form. Check it out:

Samantha has a "free gift with purchase" makeup bag full of lipsticks, but she doesn't know how many lipsticks are inside. Her friend Cheryl takes the makeup bag and magically makes the amount of lipsticks double. Then Cheryl takes three lipsticks out and gives the bag back to Samantha.

a. How many lipsticks does Samantha have now compared to what she started with?

b. And now, a new part: What if, when Cheryl gives the bag back, Samantha ends up with the same amount she started with? What number of lipsticks would Samantha have had to start out with?

Answers:

a. If Samantha started with S lipsticks, she now has $2S - 3$ lipsticks. (We worked out this solution on pp. 242–3.)

b. Well, the problem says that the *final* amount is the same as the *starting* amount, and it wants to know how much that would be, right? Let's translate that into math. So what's the starting amount? It's S. And what's the final amount? It's $2S - 3$.

Let's create an equation that *assumes* the two amounts are equal, and then we'll solve for S to see what S would need to be. Now, you might say, "How can we just assume that they are equal?" Remember: we're playing the "what if" game in these types of equations. *What if* the two amounts, S and $2S - 3$, were equal to each other?

What number would S have to equal for that to be true? Let's set them equal to each other and see what we get when we solve for S.

$$S = 2S - 3$$

If we find a value for S that makes the equation $S = 2S - 3$ true, then we've solved it!

Notice that in word problems using algebra, once you're comfortable with the math sentence you've written, solving the equation is actually the *easy* part.

Solving for S: remember, the goal is to get S by itself first, so we want all the S's on one side of the equation. (We could have rewritten the equation using boxes instead of S's, but for this problem, I'll show you what it looks like with the variables.) So let's subtract **S** from both sides, and we get:

$$S = 2S - 3$$

$$\rightarrow S - \mathbf{S} = 2S - \mathbf{S} - 3$$

(remember, two boxes minus one box equals one box!)

$$\rightarrow 0 = S - 3.$$

Now add 3 to both sides, since we still haven't gotten S by itself.

$$\rightarrow 0 + \mathbf{3} = S - 3 + \mathbf{3}$$

$$\rightarrow 3 = S$$

And look—when we get S by itself, we end up finding the answer! Whenever you get an answer like this, *always* check it in the original problem to see if it makes sense.

So, if Samantha started out with 3 lipsticks, and Cheryl doubled it, there would be 6 lipsticks. Then, when Cheryl took away 3 lipsticks, Samantha would be left with $6 - 3 = 3$ lipsticks. Yep, the starting and finishing amounts are the same, so we got it right!

Doing the Math

Write these expressions in "math language," using the variables I give you, and then solve for them. (Feel free to use boxes, if you want.) I'll do the first one for you.

1. Kelly bought 3 identical dresses. They all cost the same amount; let's call the price of each dress d. While she was shopping, she also bought a ring that cost $20.

   a. In terms of d, how much did she spend? (Assume there's no tax.)

   b. If she spent a total of $110, how much did each dress cost? In other words: solve for d.

   c. What if she spent a total of $140? Then how much would each dress have to cost?

<u>Working out the solution</u>:

a. If one dress costs d, then 3 dresses would cost $3 \times d$, or $3d$. She also spent $20 on a ring. So the *total* that she spent, in terms of d, is $3d + 20$.

b. Now we're saying the total is $110. So we can set up the equality:

$$110 = 3d + 20$$

And if we solve for d, then we will have found the value of d that makes this a true statement. See how that works? We're playing the "what if" game again: *what if* the final amount were $110—then what would d, the price of the dress, have to be in order for that equation to be true? Before moving on, make sure this all makes sense. So, to solve for d, let's get all of the numbers on one side and the variables on the other, by subtracting 20 from both sides.

$$110 - \mathbf{20} = 3d + 20 - \mathbf{20}$$
$$\rightarrow 90 = 3d$$

Now divide both sides by 3:

$$\rightarrow \frac{90}{3} = \frac{3d}{3}$$

$$\frac{\overset{30}{\cancel{90}}}{\cancel{3}} = \frac{\cancel{3}d}{\cancel{3}}$$

$$\rightarrow 30 = d$$

So, if the total was $110, then she spent $30 on each dress.

c. If the total had been $140, then we would set up the following equality (since the only thing that's changed is the amount of the total):

$$140 = 3d + 20$$

We'd do the same steps:

$$140 - \mathbf{20} = 3d + 20 - \mathbf{20}$$
$$\rightarrow 120 = 3d$$

$$\frac{\overset{40}{120}}{\cancel{3}} = \frac{\cancel{3}d}{\cancel{3}} \longrightarrow \$40 = d$$

So if the total was $140, then she must have spent $40 on each dress.

<u>Answer:</u>

a. $3d + 20$

b. $30

c. $40

2. Let's say that Brandon bought his girlfriend 4 posters for her wall. His girlfriend loves the tango, and all 4 posters have to do with tango dancing. All of the posters were the same price. Let's call the price of one tango poster p. While he was at the store, he also bought a book about filmmaking. The filmmaking book cost him $15.

a. In terms of p, how much, total, did Brandon spend at the store?

b. If Brandon spent a total of $95, how much did he spend on each poster? In other words, what is *p*?

3. I bought 5 identical boxes of chocolate for Christmas presents last year. Let's call the number of chocolates in each box *c*. While I was wrapping gifts, I ate 6 chocolates out of one of the boxes. Yeah, okay, one of the boxes was for me.

a. In terms of *c*, how many total pieces of chocolate are left?

b. What if there were now a total of 69 pieces left? How many chocolates are in each full box?

4. Lucy said to Victoria, "One year from now, I'll be twice as old as you are today." Let's call Victoria's age today *v*.

a. What is Lucy's age one year from now, expressed in terms of *v*?

b. What is Lucy's age today, expressed in terms of *v*?

c. If Lucy's age today is 15, how old is Victoria today? (in other words, solve for *v*)

(Answers on p. 293)

 *Takeaway Tips*

- Solving word problems using algebra is just like solving other word problems, in that you need to "translate" from English into math. With algebra, when you see words like *what* or *how much*, you can translate them into *variables*.

- Working with variables like *x* and *y* can take some getting used to, so don't worry if it feels strange for a while. Remember the bag of pearls (*x* = 1 bag of pearls), and refer back to chapter 20 as much as necessary!

A Final Word

I hope this book has been helpful, and that you can return to it again and again for reference and reminders. I'm proud of you for demonstrating the effort you already have in the very act of using this book to further your understanding of these math concepts. I want you to feel good and confident in your math skills, and I know you can do it!

Math isn't *easy* for anyone. It takes time and persistence to understand this stuff, so don't give up on yourself just because you might feel frustrated. Everyone feels like that sometimes—everyone. It's what you *do* about those feelings that makes you who you are.

It's in those moments when you want to give up but you *keep going anyway* that you separate yourself from the crowd and build the skills of patience and fortitude that will allow you to excel throughout your entire life—no matter what you choose as a career.

Please visit me online at: mathdoesntsuck.com. Feel free to ask questions, share comments, etc. This doesn't have to be a one-way street—I want to hear what you have to say!

Remember, math might be tough sometimes, but it *doesn't* suck—and smart is friggin' *sexy*!

Love,

Danica

Troubleshooting Guide

Where to Turn When You Don't Know What to Do!

Got troubles? Do any of these issues sound familiar?
Find your answers here!
1. "Math bores me to death."
2. "When it's time to do math, I get scared and try to avoid it."
3. "I get confused and lost during class."
4. "I think I understand something, but then I get the wrong answer in my homework."
5. "My homework is fine, but when it comes time for a test, I freeze up and can't remember anything."

Issue #1:
Math Bores Me to Death

Problem: When you sit down to do your math homework, do you have trouble focusing? Are you antsy and easily distracted?
Solution: Harness the power of pretending!

We all played make-believe at some point in our childhoods. Maybe you had an imaginary friend. Maybe you and your sister would pretend you were sleuth detectives, trying to solve a murder mystery in your backyard. Or maybe you liked to close your eyes and pretend that your bedroom was a grand castle in a foreign country, and you were queen of the land, and everyone (including your parents) had to do exactly what you told them.

Imagination can be a very powerful thing—so powerful, in fact, that it can even help get you through your math homework!

Try this: Open your math book and look at the first problem you have to do. Read it to yourself in your head, and interject all sorts of enthusiastic words into it. Seriously, I know this is wacky and totally uncool, but no one can hear what goes on inside your head. (Thank goodness, right?)

For example, if the problem is $\frac{1}{4} + \frac{3}{4} = ?$ Think to yourself, "Ah, $\frac{1}{4}$! I love $\frac{1}{4}$! What a great fraction! And I can't wait to add it to $\frac{3}{4}$. Yay! Let's see, how should I do this? Well, since the denominators are the same, I can just add across the top. Great! So the answer is $\frac{4}{4}$. But wait! That doesn't look reduced; I have to reduce it first. I mean, I *get* to reduce it, because I love reducing fractions almost as much as I love adding them together! Yippee!"

Make sure that the little voice saying this dialogue is *really* peppy—like a cheerleader. And no matter how many times your own real voice tries to interrupt the "enthusiastic" voice, just keep saying, "Yay! I can't wait to do the next problem! There's nothing else I'd rather be doing!" Let the little voice steamroll over any other thoughts that might try to come in.

I'm sure this all sounds very strange. But you know what? *It works.*

I mean, don't you hate it when you've been staring at your homework for, like, an hour without getting much done because you just can't mentally stick to it? And then you've wasted an hour?

Look, the more outrageously positive your thoughts, the better. The more you can make yourself *believe* the pretend thoughts, the better. If you can crack yourself up, *even* better.

You'll be amazed at how much *pretending* to love your math homework can help you focus, and speed things along. It can even help during exams, if you're having a hard time focusing—just make sure you think these "cheerleader" thoughts *inside* your head and not out loud!

Danica's Diary
FAKE IT TILL YOU MAKE IT!

When I was studying for the SATs, my least favorite part was the "reading comprehension" section. In this section, you have to read these big paragraphs about, I don't know, ant farms and other subjects that seemed really boring to me. Then you have to answer detailed questions about what you read.

Every time I did a practice test, I felt my mind drifting midway through the stories; I just couldn't find the energy to focus on them. And as a result, when I got to the end of the stories, I couldn't remember much about them—so I had a really hard time answering the questions. As much as I tried to concentrate, I kept thinking, "I have no interest in ant farms. Why should I read this?" Even though, technically, I knew the answer: because it would help me do well on my SATs, which would help me get into a good college. Yes, my entire future depended on me figuring out how to focus on stories about things like ant farms.

Then my SAT tutor told me about this "cheerleading" idea. She said, "I know many of these stories are not interesting to you, but you need to read them closely, paying attention to their details, so here's what I suggest: as you're reading them, pretend like they're the most fascinating thing you've ever read."

I thought she was insane. But I tried it. I sat there reading these boring paragraphs and thinking pretend thoughts like, "Wow. I just love ant farms. How many cubic inches do the ants burrow on an average per day? Fascinating! Tell me more about the different types of ants!"

The thoughts were totally stupid—but sort of funny. I soon found myself becoming slightly amused by the stories and actually remembering more of their details afterward.

In the end, adopting the "cheerleader" approach helped me to improve my reading comprehension scores quite a bit. All because of the power of pretending.

Issue #2: When It's Time to Do Math, I Get Scared and Try to Avoid It

Problem: Do you feel queasy at the thought of doing math? Do you think math is scarier than your cousin's freaky Halloween mask? **Solution**: Get competitive, and go from fear to strength!

It's okay to be scared of math sometimes. We all know that feeling of dread as we delve into a new math subject: "What if I can't do it? What if I fail?"

These are totally normal feelings. Everyone is afraid of failure. Did you hear me? *Everyone*—kids, adults, you name it. Some may not show it, but inside we all experience the same types of feelings toward the things that challenge us.

Sometimes fear can feel like a pit in your stomach; other times it can feel like a dizzying sensation, running throughout your entire body. Some people feel this way before making a speech in front of their class or calling a guy they like—or when faced with a confusing math problem that makes them feel like a failure for not being able to immediately understand how to solve it.

The difference between those who succeed and those who don't is *not* whether or not they feel scared. It's how they *handle* the fear. This is true in math and in many other areas of life as well. The key is to *not give in* to these feelings—to not let them control us.

"But how can I do that?" Simple: Get in touch with your competitive, aggressive side! Refuse to be beaten. Every time we face a fear, every time we are afraid of a new challenge but then push through anyway, we get *stronger*—mentally and emotionally. Did you hear me? We actually get <u>stronger</u>. When you're feeling scared or intimidated by math, recognize it as an opportunity to *practice* getting stronger.

The great Eleanor Roosevelt* once said, "Do one thing every day that scares you." Now you know why. Practicing facing intimidation square in the face and saying, "I'm not going to let you beat me," will make you better at math, and it will also make you a more successful person in all areas of your life.

Issue #3: I Get Confused and Lost During Class

Problem: Do you sit in class *trying* to understand your teacher, but none of it makes sense? Are you embarrassed to raise your hand when you have a question?
Solution 1: Don't be shy—be bold!
Solution 2: Read ahead in your textbook. (You probably won't understand what you read, but that's okay—you're not supposed to!)

Solution 1: Don't Be Embarrassed to Ask Questions!

Teachers are there to teach you; that's their job. You can ask them questions before, during, or after class. Give it a try! They won't bite. Well, most of them, anyway.

During Class

If you're confused by something during class, there's an almost 100% chance that *at least* a few other people in your class are confused by the same thing. And if you have a question, there's a *huge* chance that several other people in the class *want* to ask the same question but are too shy to raise their hand.

When you raise your hand and ask the same question that is burning in the minds of some of the silent students around you, *they will be secretly relieved that you did*. You'll probably never know which students they were, but trust me, they're out there.

The reason most of us shy away from asking questions is because

---

* *Eleanor Roosevelt (1884–1962): First lady to U.S. president Franklin Delano Roosevelt. When he took office in 1933, she revolutionized the position of "first lady" by holding her own press conferences, hosting regular radio broadcasts, and writing a daily newspaper column. After FDR's death in 1945, Eleanor led a vibrant career as a U.S. delegate to the United Nations and was awarded the first United Nations Human Rights Prize. She was never considered physically attractive, and at the age of 14, she wrote the touching words, ". . . no matter how plain a woman may be, if truth & loyalty are stamped upon her face, all will be attracted to her."*

we are afraid of looking dumb. We think, "What if I'm the only one who doesn't get it? I should understand this, and it's my fault that I don't, so I should just stay quiet and hope I catch on later. I don't want to be embarrassed in front of my classmates."

Listen to me, *even if,* when you ask a question, someone in the class thinks, "What a moron. I can't believe she asked such a stupid question"—and this is *very* unlikely—an hour later, they will most likely have completely forgotten that you ever raised your hand and will instead be busy thinking about themselves (and probably silently worrying about what others are thinking of *them*).

So ask your questions in class. You'll be glad you did, and so will your classmates.

QUICK NOTE! Did you know that girls are a lot less likely than boys to raise their hand in class to ask questions? It's true. Studies have found that this is because we girls are taught to be "polite." (I'm sure you'll agree that, in general, the girls you know are more polite than the boys. Girls just mature faster, that's all!) Being polite is wonderful in many situations, but know this: There is nothing impolite about asking questions!

Didja Know?

Being confused about a new concept (in math and other subjects) is part of the natural human learning process. In other words, you're *supposed* to be confused when you're first learning something—and you're *supposed* to have questions. So when your teacher asks, "Any questions?" don't be shy. Be bold!

Also, teachers don't always know how they "sound" when they are teaching; sometimes they make mistakes or forget an important part of the explanation. They may not have realized that they left something out until *you* ask your question.

After Class

If you ask a question during class but don't understand the answer, or the teacher's explanation is just plain unclear, it's perfectly okay to approach the teacher after class and ask him or her again. If the teacher doesn't seem to want to help you—or it seems like she or he may not understand the math him- or herself (this happens sometimes, believe it or not)—then it's time to look outside for help. There may be other math teachers at your school who teach the same grade level and who would be willing to help you, or you might find a tutoring program at school. Even if you just have one or two questions, you'll find that when you reach out for help, good teachers will usually be happy to explain stuff to you.

Think about it from their point of view. Especially in math, teachers usually encounter students who don't care about learning. So it's *refreshing* when a student goes out of her way to approach the teacher and seek out their help.

So don't worry. Just be respectful when you ask a teacher for help, and you'll be fine. You might even be surprised at how delighted she is that you asked.

If All Else Fails . . .

If you can't find any teachers who are able to help you, have no fear: there are countless other resources that can explain the same math concepts you're learning in your classroom.

Remember, math is universal. Although there can be slight differences in a teacher or book's approach to any given problem, whether you get help from a tutor, this book, a math tutoring website, or your brilliant IM buddy from France, we'll all be explaining the *same math concepts*.

On my website, www.mathdoesntsuck.com, I've included a "Smart Girl's Resource Guide" to help you locate some of these resources.

Solution 2: Read Ahead

There is no reason that you can't *read ahead* in your book the night before you are going to learn a section—this is a great way to help you follow along in class.

You're probably thinking, "But if I don't understand these things when my teacher teaches them to me, how am I supposed to understand them from reading the book?"

You're not. In fact, there's absolutely *no* pressure on you to understand it at all—that's the beauty of reading ahead. You can just skim the words and get a "feeling" for what's coming up next. As you read, you can be like, "Wow, what's a hypotenuse? *Hmm*, looks like a little line on a triangle thingy. Neato! Whatever!" Oddly enough, just seeing the new concepts in front of you, even briefly, will actually help you to understand them the next day in class. The best part is, since you don't have to learn anything when you're skimming like this, it only takes a few minutes!

Issue #4: I Think I Understand Something, but then I Get the Wrong Answer in My Homework

Problem: Always making careless mistakes on your homework? Can't seem to get the answer right, even though you thought you understood the topic?
Solution 1: Read the directions—*closely*.
Solution 2: Stop racing through your homework.
Solution 3: Don't worry about saving paper.

Have you ever seen a woman carrying a little dog in her shoulder bag? Some stores actually sell "dog purses," specifically designed to hold small dogs and puppies. Maybe one of these stores has a sign in its window, saying:

Do you see anything wrong with this sign? Did you read it carefully?

Read it again.

If you noticed that something was wrong with the sign the first time you read it, then you are among something like 1% of the population.

When *most* people read this phrase, they think it says, "Puppy in the purse," when it really says, "Puppy in the the purse." Most people totally miss the second *the*.

Why does this happen? Because we don't expect the second *the* to be there. In fact, we expect it *not* to be there; our brains jump ahead and *interpret* what we're reading, instead of reading what's actually on the page.

This happens while we're reading math instructions, too. When doing your math homework or taking a test, always make sure to read the directions—closely. Even if you *think* you've read the directions (and the problem), unless you've paid attention to each and every word, you really haven't read them.

Solution 1: Read the Directions!

This solution may seem silly and obvious to you, but you wouldn't believe the mind's ability to trick us into thinking we're reading what we *expect* to see there, instead of what's really there. It happens in more than just math class. How did you fare with the puppy sign? 'Nuf said.

Solution 2: Stop Racing Through Your Homework

I know, I know. Nobody wants to spend more time on their math homework than they absolutely have to; most people take whatever shortcuts they can in order to finish it as fast as humanly possible.

We all like shortcuts. Why not? They make life more efficient, right? Sure, sometimes. But I'm going to let you in on a little secret: you can avoid a lot of frustration—and most of your "careless errors"—by *not* taking so many shortcuts during your math homework.

You know how sometimes your teacher says, "show your work"? There's actually a good reason for that. Most people skip steps in an attempt to get their work done faster and then make mistakes they can't even catch, because they didn't write them down!

I hate that: going over the page again and again, trying to figure out what went wrong. You can see how this actually *adds* time to the whole process. If, instead, you write down all the steps along the way, you're less likely to make a mistake in the first place—and if you do, it's much easier to immediately see what went wrong.

Listen, I know these tips sound like a waste of time. I've been exactly where you are. But listen to what I'm saying: they may add a tiny bit of time up front but will save you countless hours of frustration later. I *promise*.

And when you start getting *more* answers right *more* of the time, you'll find that you actually understand math much better than you thought you did!

Solution 3: Don't Worry About Saving Paper

When you are doing your math homework, *take up space on the page*.

It's wonderful to care about the environment, and to want to save paper, but *not* while you're doing your math homework. Do you hear me? *Not* while you're doing your math homework.

Look, I totally understand the desire to conserve paper. When I was in middle school, I would always try to cram as many problems onto a homework sheet as I possibly could. It seemed the most "efficient" way to go, and I felt good about having fewer pages to turn in.

For example, if it looked like *almost* all of the problems were going to fit onto one page, I would do whatever I had to in order to get *all* of the problems to fit onto one page. I'd squeeze 'em in there any which way I could. I would think, "Phew! I finished the homework *and* I got it all on one page." Cramming it all in felt like extra bonus points or something—an accomplishment all of its own.

Does this sound like you?

If so, you have to admit that this kind of thinking is kind of silly. The truth is, there is nothing *better* about using less paper for your homework. And there are plenty of bad things about it!

First of all, it takes your focus away from the math problem you are doing and puts it on your "mission" to squeeze it all in. Any time your focus is split like that, you're more likely to make mistakes. I know I did.

Second, by cramming in the last few problems, your writing has to get smaller, which makes it harder to read what you've done—and easier to make mistakes.

Third, if you're trying to fit it all onto fewer sheets of paper, you're more likely to *skip steps* in your solution in an attempt to take up less space. And as you know from above, skipping steps in your solution *always* increases the chances that you'll make mistakes, and it also makes it difficult to find the mistakes later.

My middle school goddaughter Tori recently told me about some steps she skips in fraction division in order to save pencil lead. Yes, she was trying to save *pencil lead*! If you think about it, though, that's not much crazier than trying to save an extra piece of paper (which, let's face it, can get recycled later on).

Where does this kind of thinking come from? Is it environmentalism? Is it wanting to feel more efficient? For me, I think it was a little of both—but in the grand scheme of things, your efficiency will actually go *up* when you start making fewer mistakes.

Which one looks like you?

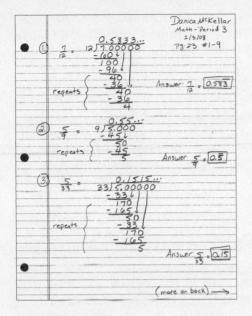

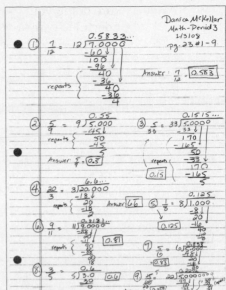

So go ahead—spread out! Take up *as much space on the page* as you need to. And if you end up with just one problem all by itself on a page? Well, then I guess you know you're getting the hang of it!

Neat and Pretty

Many of us gals like our homework to look neat and pretty—we don't like to have a bunch of chicken scratch all over it. In many ways, this is a good thing! A neater homework sheet will make it easier to find mistakes.

The problem is, in trying to be neat and pretty we tend to do more of the thinking in our head and less of it on the page, which increases the chances that we'll make mistakes. The more you *write down* as you're working out a problem, the easier it is to understand what you're doing, while you're doing it. Not to mention, it usually ends up being much *faster* to solve problems this way: even though this approach may seem like more work at the time, it usually enables you to think more clearly, which allows you to find the solution faster.

But, it's messy.

What's a girl to do? How can we write down all our messy thoughts and possible solutions, without having messy homework?

Here's what I recommend: Keep a separate "scratch sheet" next to you as you do your homework. Work the problems out on that scratch sheet, writing down all your thoughts for how to solve it, and then only write down the basic steps of the solution on your neat and pretty homework page.

Some teachers will say, "Show all your work." In that case, you can copy some of your messy, half-confused brainstorming thoughts from your scratch sheet to your neat and pretty homework page, but you can still be selective about which things you write down.

After you finish your homework, you'll see that some of the scratch sheets will have a bunch of nonsense on them; just cross out the solutions you tried that didn't work. Then, *save your scratch sheets* until after final exams. You don't have to keep them in your binder; it's probably stuffed already. You can keep them somewhere else, maybe at home, in a separate folder.

At some point while you're studying for your test and reviewing your homework, you'll say to yourself, "How did I get that answer?!" And you'll be so happy if you can refer back to your scratch sheets, because you'll *see* exactly how you arrived at a certain answer.

Also, in some schools, teachers take a long time to grade and return homework assignments, so the scratch sheets are extra helpful when you don't even have your homework back, but you need to review the material for an upcoming test.

You'll be, like, "Thank you, darlin', for saving these scratch sheets. You're a genius." Yes, at that point (sometime in the future) you'll be talking to yourself in the past. Creepy, isn't it?

Here's an example of a problem done on a scratch sheet and then translated to the "neat and pretty" homework page.

Example

You almost bought a scarf for $28, but then you went to a second store and saw the same scarf for $24, plus it was even marked $\frac{1}{3}$ off. How much more would you have spent if you'd bought it at the first store?

Scratch Sheet:

First store's price: $28

Second store's price: $24 – with 1/3 off

First: Find second store's price –

What is $\frac{1}{3}$ of 24? → $\frac{1}{3} \times 24$

→ $\frac{1}{3} \times 24 = \frac{1}{3} \times \frac{24}{1} = \frac{24}{3} = \frac{\cancel{24}^8}{\cancel{3}_1} = \frac{8}{1} = \$8$

✱ So, $\frac{1}{3}$ off of $24 = \$24 - \$8 = \$\underline{16}$: second store's price.

Compare to first store's price: $28 - $16 = $12

Answer: I would have spent $\boxed{\$12}$ more at the first store.

Neat and Pretty Homework Page:

First store's price: $28
Second store's price: $24, with 1/3 off.

$\frac{1}{3}$ of 24 = $\frac{1}{3} \times 24$ = 8

So: $\frac{1}{3}$ off of 24 = 24 - 8 = 16. $28 - $16 = $12

Answer: $12

It might seem like I'm writing too much down on the scratch sheet. That's a good thing! You can never write too much down on a scratch sheet . . . even doodles are okay.

Issue #5: My Homework Is Fine, but When It Comes Time for a Test, I Freeze Up and Can't Remember Anything

Problem: Do you get a prickly feeling on the back of your neck when your teacher passes out a math test? Does your mind suddenly feel blank while staring at an exam?

Solution 1: Skim the whole test, then do the easy problems first.

Solution 2: Exercise "mind over matter."

Solution 1: Show the Test Who's Boss—by Skimming It First

Have you ever been handed a math test, taken one look at the first problem (which might as well have been written in Greek, it looked so foreign), and felt your stomach drop? It's one of the worst feelings in the world—and one most of us know too well.

The solution? Take control of the test by first skimming the whole thing. This is the "let's get acquainted" phase. Say, "Hello, Test, let's get to know you a little." It's kind of like turning the lights on in a house before walking in. Would you rather slowly walk down dark, scary hallways—or, instead, turn all the lights on and take a quick, friendly jog through the place so there won't be any surprises?

When you are skimming, lightly circle the problems that look the most familiar and "doable." Don't do any of them yet, though; just keep skimming until you've gotten to the end. By then, you'll have a good sense of the beast in front of you.

Once you've gone through the whole test, go back and *do the easy ones first*—there is *no* reason why you should do the first problems first. Show your test who's boss by *choosing* which problems to do first. After all, who's in control here—some little test paper, or *you*?

Solution 2: Mind Over Matter

When we are afraid of something, our bodies respond with a real, physical reaction—it's called the "fight or flight" response.

You see, many years ago, when we were all cavemen, "fear" usually meant that we were being faced with a large animal that wanted to eat us. And when we were afraid, our bodies would release a hormone called adrenaline into our system, which would make our hearts pump faster and prepare us for "fight or flight"—that is, either to fight the beast or run away from it!

Unfortunately, adrenaline doesn't help much when we are trying to calm down our minds enough to relax and think clearly on a math test. In fact, this is the main difference between homework and tests—it's not so much that you can't refer back to the chapter for help, but that there is the added psychological stress and pressure of having to "perform" on cue (which, as we know, can sometimes feel as scary as coming face-to-face with a wild jungle beast!).

So how can we calm ourselves down? The first step is simply to *know that we can*—that it's all in our heads.

Danica's Diary

BIKINI WAX BLISS (I WISH!)

Ever heard of a bikini wax? They can be pretty painful. They tell you to "relax" and, sure, you try to—but the hot wax and gauze usually make this difficult!

I'll never forget my first waxing. Sure enough, every time I tensed up, it hurt more. A lot more. It's human nature that when you think to yourself, "Don't tense up," all you want to do is tense up!

But after a few more sessions and a lot more pain, I found that I *could* calm myself down. The mind's power over the body is incredible. All I had to do was think about roses or rainbows or fluffy clouds, and it didn't hurt as much! Sounds wacky, I know, but I'm telling you—it actually works!*

How does this relate to math tests?

Once when I was in college, I had two major final exams in mathematics on the same day—they were both three hours each! To make things worse, I learned that a friend of mine was going to be finding out some really important (and possibly bad) news that day. I didn't know how I would make it through the day. I was so worried about my friend and so nervous about taking the two math tests.

That morning, my alarm radio went off, and a really peaceful song was playing; one that I'd never heard before. I remember that this feeling of calm came over me, and somehow, I knew that I'd be able to get through the day—that I was going to be okay.

Throughout both exams, whenever I'd get scared for my friend or nervous about the test, I would close my eyes for a moment and think about the song. I imagined that the singer was singing directly to me. I felt so taken care of, so nurtured.

....

* This is still so hard for me to believe that sometimes I'll tense up just to "test" it—and sure enough it kills. In the next moment I'll think about fluffy clouds again, and it hardly hurts at all. I'm not even kidding!

The amazing thing was, as I felt my body become more relaxed physically, my mind would relax as well, enabling me to think more clearly and solve the problems on the test. (And yes, I skimmed the tests at the beginning of each and did the easy problems first, which also helped to alleviate some of my nerves!)

Thankfully, that night, I also found out that my friend was okay.

Do you have an image, a memory, or a song that makes you feel really peaceful and taken care of? Something that you can use as a tool to keep your body and mind relaxed in the midst of stressful situations? Write it down on a sticky note and put it somewhere where you'll see it when you're about to take a math test. This stuff works.

Overcoming fear—in math and other areas of life—is very much psychological. You can "psych" yourself out of being afraid—it just takes time to build the tools that work for you, and over the years you'll be able to see your progress (especially if you keep a journal). Not everyone works to improve themselves like this—but you can choose to, and believe me, you'll find it a worthy investment of your time. Good luck, and I hope this book has provided you with some tools that will serve you well in math and beyond!

The Smart Girl's Resource Guide

Here are some resources for help and inspiration!

You can also visit this book's website, www.mathdoesntsuck.com, for a more up-to-date list—because as you know, things change pretty fast online!

Online tutoring:

mathforum.org/dr.math/

Not sure how quickly they answer people's questions, but the FAQ section of this site is very interesting!

More explanations for topics that still confuse you:

www.homeschoolmath.net/math_resources_3.php
www.visualfractions.com/
math.com/homeworkhelp/PreAlgebra.html

Here's a very funny site with some great advice for math students:

www.mathpower.com/

Once you've finished your homework, here are some other great sites to visit:

www.sallyrideclub.com/member_home.do
expandingyourhorizons.org/GirlsArea/Intro.html

These two sites can tell you about some great events in your area, where you can meet other girls your age and learn about math and science. They even have games and competitions!

Math-A-Thon:

www.mathathon.org

You might also consider joining Math-A-Thon, a great way to help kids with cancer. Math-A-Thon is a program through St. Jude Children's Research Hospital where you raise money for cancer research by doing math problems from home and getting sponsored by friends and family, like a marathon! You can also win prizes, depending on how much money you raise. I've also posted some solved math problems on that site, for grades K–8, so be sure to check it out!

And finally, just a few sites where you can meet and hang out with other smart girls like you:

www.girlstart.org/
www.girlsinc.org/gc

Multiplication Tables

While I recommend memorizing your multiplication tables, here they are for your reference, just in case!

A Quick Trick for Remembering Your Nines

What's 9 × 6? Hold out all 10 fingers and lower the 6th finger. There are 5 fingers to the left and 4 fingers on the right—and the answer is 54! This works with all the 9s from 9 × 1 through 9 × 9. Try it!

The Multiplication Facts

1 × 1 = 1	4 × 1 = 4	7 × 1 = 7	10 × 1 = 10
1 × 2 = 2	4 × 2 = 8	7 × 2 = 14	10 × 2 = 20
1 × 3 = 3	4 × 3 = 12	7 × 3 = 21	10 × 3 = 30
1 × 4 = 4	4 × 4 = 16	7 × 4 = 28	10 × 4 = 40
1 × 5 = 5	4 × 5 = 20	7 × 5 = 35	10 × 5 = 50
1 × 6 = 6	4 × 6 = 24	7 × 6 = 42	10 × 6 = 60
1 × 7 = 7	4 × 7 = 28	7 × 7 = 49	10 × 7 = 70
1 × 8 = 8	4 × 8 = 32	7 × 8 = 56	10 × 8 = 80
1 × 9 = 9	4 × 9 = 36	7 × 9 = 63	10 × 9 = 90
1 × 10 = 10	4 × 10 = 40	7 × 10 = 70	10 × 10 = 100
1 × 11 = 11	4 × 11 = 44	7 × 11 = 77	10 × 11 = 110
1 × 12 = 12	4 × 12 = 48	7 × 12 = 84	10 × 12 = 120
2 × 1 = 2	5 × 1 = 5	8 × 1 = 8	11 × 1 = 11
2 × 2 = 4	5 × 2 = 10	8 × 2 = 16	11 × 2 = 22
2 × 3 = 6	5 × 3 = 15	8 × 3 = 24	11 × 3 = 33
2 × 4 = 8	5 × 4 = 20	8 × 4 = 32	11 × 4 = 44
2 × 5 = 10	5 × 5 = 25	8 × 5 = 40	11 × 5 = 55
2 × 6 = 12	5 × 6 = 30	8 × 6 = 48	11 × 6 = 66
2 × 7 = 14	5 × 7 = 35	8 × 7 = 56	11 × 7 = 77
2 × 8 = 16	5 × 8 = 40	8 × 8 = 64	11 × 8 = 88
2 × 9 = 18	5 × 9 = 45	8 × 9 = 72	11 × 9 = 99
2 × 10 = 20	5 × 10 = 50	8 × 10 = 80	11 × 10 = 110
2 × 11 = 22	5 × 11 = 55	8 × 11 = 88	11 × 11 = 121
2 × 12 = 24	5 × 12 = 60	8 × 12 = 96	11 × 12 = 132
3 × 1 = 3	6 × 1 = 6	9 × 1 = 9	12 × 1 = 12
3 × 2 = 6	6 × 2 = 12	9 × 2 = 18	12 × 2 = 24
3 × 3 = 9	6 × 3 = 18	9 × 3 = 27	12 × 3 = 36
3 × 4 = 12	6 × 4 = 24	9 × 4 = 36	12 × 4 = 48
3 × 5 = 15	6 × 5 = 30	9 × 5 = 45	12 × 5 = 60
3 × 6 = 18	6 × 6 = 36	9 × 6 = 54	12 × 6 = 72
3 × 7 = 21	6 × 7 = 42	9 × 7 = 63	12 × 7 = 84
3 × 8 = 24	6 × 8 = 48	9 × 8 = 72	12 × 8 = 96
3 × 9 = 27	6 × 9 = 54	9 × 9 = 81	12 × 9 = 108
3 × 10 = 30	6 × 10 = 60	9 × 10 = 90	12 × 10 = 120
3 × 11 = 33	6 × 11 = 66	9 × 11 = 99	12 × 11 = 132
3 × 12 = 36	6 × 12 = 72	9 × 12 = 108	12 × 12 = 144

Answer Key

For the fully explained solutions, visit the "Solution Guides" page at mathdoesntsuck.com.

Doing the Math from p. 8

2. $15 = 3 \times 5$

3. $75 = 3 \times 5 \times 5$

4. $100 = 2 \times 2 \times 5 \times 5$

5. $48 = 2 \times 2 \times 2 \times 2 \times 3$

Doing the Math from p. 10

2.

3.

4.

Doing the Math from p. 17

2. 14

3. 10

Doing the Math from pp. 19–20

    2.   45

    3.   25

Doing the Math from p. 23

    2.   8

    3.   3

    4.   18

Doing the Math from p. 28

    2.   5: 5, 10, 15, 20, 25, 30, 35, 40, 45, 50

    3.   7: 7, 14, 21, 28, 35, 42, 49, 56, 63, 70

    4.   12: 12, 24, 36, 48, 60, 72, 84, 96, 108, 120

Doing the Math from p. 32

    2.

    9:   9, 18, 27, <u>36</u>, 45, 54, 63

    12:   12, 24, <u>36</u>, 48, 60, 72

Answer: 36

    3.

    6:   6, 12, 18, 24, 30, 36, <u>42</u>, 48

    7:   7, 14, 21, 28, 35, <u>42</u>, 49

Answer: 42

    4.

    4:   4, 8, 12, <u>16</u>, 20

    16:   <u>16</u>, 32, 48

Answer: 16

5.

9:   9, 18, 27, 36, <u>45</u>, 54

15:   15, 30, <u>45</u>, 60

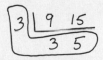

Answer: 45

Doing the Math from p. 40

2. $\frac{2}{5}$

3. $\frac{7}{73}$

4. $\frac{4}{7}$

Doing the Math from pp. 43–4

2. $2\frac{2}{3}$

3. $1\frac{1}{5}$

4. $3\frac{1}{4}$

Doing the Math from p. 46

2. $\frac{5}{2}$

3. $\frac{20}{3}$

4. $\frac{8}{5}$

Doing the Math from p. 48

2. $\frac{6}{1} = 6$

3. $\frac{1}{1} = 1$

4. $\frac{141}{1} = 141$

Doing the Math from p. 53

2. $\frac{3}{2}$

3. $\frac{10}{21}$

Doing the Math from p. 56

2. $\frac{3}{8}$

3. $\frac{2}{5}$

4. $\frac{296}{19}$

5. $\frac{1}{9}$

Doing the Math from p. 59

2. $\frac{8}{15}$

3. $\frac{16}{25}$

4. $\frac{1}{6}$

Doing the Math from p. 66

2. $\frac{2}{4}, \frac{3}{6}, \frac{10}{20}$

3. $\frac{8}{6}, \frac{12}{9}, \frac{40}{30}$

4. $\frac{10}{2}, \frac{15}{3}, \frac{50}{10}$

Doing the Math from pp. 72–3

2. $\frac{2}{3}$

3. $\frac{73}{84}$

4. $\frac{4}{5}$

Doing the Math from pp. 77–8

2. $\frac{3}{4} < \frac{4}{5}$

3. $2\frac{1}{3} = \frac{21}{9}$

4. $\frac{5}{11} < \frac{1}{2}$

Doing the Math from pp. 81–2

2. $\frac{20}{14} < \frac{90}{60}$

3. $\frac{1}{21} > \frac{1}{22}$

4. $\frac{100}{3} > \frac{3}{100}$

5. $\frac{17}{51} = \frac{1}{3}$

6. $\frac{2}{6} < \frac{3}{8}$

Doing the Math from p. 87

2. $\frac{8}{3}$ or $2\frac{2}{3}$

3. different denominator

4. $\frac{1}{2}$

5. 2

Doing the Math from pp. 92–3

2. $\frac{4}{9}$

3. $\frac{1}{36}$

4. $\frac{25}{72}$

5. $\frac{7}{6}$ or $1\frac{1}{6}$

Doing the Math from p. 101

2. 3

3. $\frac{1}{2}$

4. $\frac{3}{4}$

Doing the Math from pp. 108–9

2. 4

3. $\frac{9}{2}$ or $4\frac{1}{2}$

4. $\frac{1}{2}$

Doing the Math from p. 119

2. $0.8888 < 0.891$

3. $0.45 > 0.1999$

4. $56.11 > 6.889$

5. $0.1112 < 0.1211$

Doing the Math from p. 121

2. 24.401

3. 73.591

4. 5.61

5. 27.5

Doing the Math from p. 125

2. 0.18

3. 9.1

4. Yes! They'd charge $2.89 (with taxes, they always round up on the last penny).

Doing the Math from p. 127

2. Divisor = 3, Dividend = 63, Quotient = 21

3. $4\overline{)32}^{\,8}$ Divisor = 4, Dividend = 32, Quotient = 8.

4. $5\overline{)80}^{\,16}$ Divisor = 5, Dividend = 80, Quotient = 16.

5. $8\overline{)72}^{\,9}$ Divisor = 8, Dividend = 72, Quotient = 9.

6. $2\overline{)10}^{\,5}$ Divisor = 2, Dividend = 10, Quotient = 5.

Doing the Math from p. 130

2. 0.13

3. 0.013

4. 13.4

Doing the Math from pp. 133–4

2. 5

3. 7

4. 7

5. 0.5

Doing the Math from p. 140

2. 0.4

3. 0.125

4. 1.5

Doing the Math from p. 142

2. 1.2

3. 2.75

4. 3.5

Doing the Math from p. 146

2. $0.2\overline{6}$

3. 0.4

4. $0.\overline{69}$

5. $1.\overline{1}$

Doing the Math from p. 156

2. $\frac{4}{5}$

3. $\frac{8}{9}$

4. $\frac{3}{2}$ or $1\frac{1}{2}$

5. $\frac{14}{9}$ or $1\frac{5}{9}$

Doing the Math from p. 167

2. 0.05

3. 75%

4. 5

5. 0.0009

6. 144%

7. 0.005

Doing the Math from p. 171

2. $\frac{1}{4}$

3. $\frac{1}{500}$

4. $\$45$

Doing the Math from p. 173

2. 50%

3. 150%

4. 400%

Doing the Math from p. 180

2. $0.08, 8\%, \frac{2}{25}$

3. $5, 500\%, \frac{5}{1}$

4. $0.75, 75\%, \frac{3}{4}$

5. 0.025, 2.5%, $\frac{1}{40}$

6. 0.004, 0.4%, $\frac{1}{250}$

Doing the Math from pp. 181–2

2. 16%, $\frac{1}{6}$, 0.19

3. $\frac{7}{4}$, $1\frac{4}{5}$, 200%

4. $\frac{8}{9}$, 0.889, 89%

Doing the Math from p. 188

2. 0.6×10

3. $\frac{1}{3} \times 30$

4. $0.16 \times \frac{1}{3} \times 600$

5. $10 - (0.6 \times 10)$ or 0.4×10

Doing the Math from p. 191

2. $21

3. $80

4. $1.50 per magazine, part 2: For the full year, we'd pay $18.

Doing the Math from p. 198

2. "4 to 3" or 4:3 or $\frac{4}{3}$

3. "3 to 1" or 3:1 or $\frac{3}{1}$

4. "10 to 1" or 10:1 or $\frac{10}{1}$

5. "5 to 3" or 5:3 or $\frac{5}{3}$

Doing the Math from pp. 206–7

2. $\frac{6.4 \text{ miles}}{1 \text{ day}} = 6.4$ miles per day

3. $\frac{5 \text{ kids}}{7 \text{ cars}} = 5$ kids to every 7 cars

4. $\frac{\$2.50}{1 \text{ bottle}} = \$2.50$ per bottled water

5. $\frac{\$0.90}{1 \text{ ft.}} = \$0.90$ per foot

Doing the Math from p. 219

2. $m = 24$

3. $m = 12$

4. $m = 20$

Doing the Math from pp. 224–5

2. It will take 12 minutes and 30 seconds to finish the reading assignment. (That is, if you keep reading at *exactly* the same pace!)

3. You should use $\frac{1}{4}$ tsp. in the smaller version of the recipe.

4. $1\frac{1}{4}$ bags of puppy food

Doing the Math from p. 234

2. 6 miles $\approx$ 9.66 kilometers

3. 5 feet $\approx$ 1.525 meters

4. 42 purses = 140 bottles of nail polish

5. 36 pints = 4.5 gallons

Doing the Math from pp. 243–4

2. $x + 3$

3. $m - 7$

4. $2c - 3$

Doing the Math from p. 254

2. $x = 18$

3. $x = 2$

4. $x = 10$

5. $x = 5$

6. $x = 1$

Doing the Math from pp. 261–63

2a. $4p + 15$

2b. Each poster costs $20.

3a. $5c - 6$

3b. There are 15 chocolates in each full box.

4a. Lucy's age one year from now will be $2v$.

4b. Lucy's age today is $2v - 1$.

4c. Today, Victoria is 8 years old: $v = 8$.

Index

About the Author

Best known for her roles as Winnie Cooper on *The Wonder Years* and Elsie Snuffin on *The West Wing*, Danica McKellar is also an internationally recognized mathematician and advocate for math education.

Upon its release, *Math Doesn't Suck* became a national bestseller. Danica made headlines, and was named "Person of the Week" by *ABC World News with Charles Gibson*. She has been honored in Britain's esteemed *Journal of Physics* and the *New York Times* for her work in mathematics, most notably for her role as coauthor of a groundbreaking mathematical physics theorem that bears her name (The Chayes-McKellar-Winn Theorem).

A summa cum laude graduate of UCLA with a degree in mathematics, McKellar's passion for promoting girls' math education earned her an invitation to speak before Congress on the importance of women in math and science. Amidst her busy acting schedule, Danica continues to make math education a priority, as a featured guest and speaker at mathematics conferences nationwide.

McKellar is also a spokesperson for the Math-A-Thon program at St. Jude Children's Research Hospital, which raises millions of dollars every month both for cancer research and to provide free care for young cancer patients.

McKellar lives in Los Angeles, California, and this is her first book. Her second book, *Kiss My Math: Showing Pre-Algebra Who's Boss*, hits the shelves in fall 2008.